ELECTRICAL
MOTOR CONTROLS
Automated Industrial Systems
MANUAL

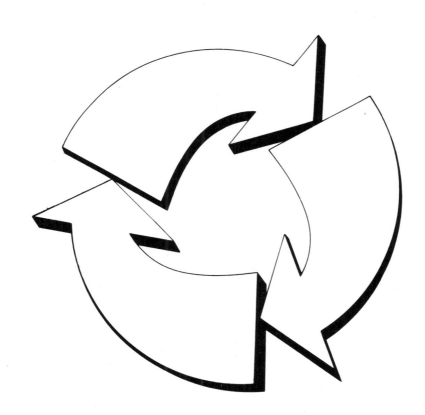

AMERICAN TECHNICAL PUBLISHERS, INC.
HOMEWOOD, ILLINOIS 60430

Glen Mazur
Jennifer Sparlin

1 2 3 4 5 6 7 8 9 - 92 - 9 8 7 6 5 4 3 2 1

Printed in the United States of America

ISBN 0-8269-1668-6

CONTENTS

INTRODUCTION

Electrical Motor Controls Manual is a hands-on, applications-filled, activities-oriented manual covering electrical control devices. *Electrical Motor Controls Manual* may be used independently. For more in-depth study, use *Electrical Motor Controls*, 3rd edition with this manual.

Technical data is provided through application sheets that show the proper use, sizing, and connection of control devices. Activities provide preparation for proper ordering, installation, maintenance, and troubleshooting of control devices and circuits. Manufacturing data is presented as it appears in service manuals used by industrial electricians. A comprehensive Appendix provides useful information in an easy-to-find format.

Activity questions include identification, completion, calculations, short answer, and illustrated answers. Always record answers in the space(s) provided. The following are examples of activity questions. All answers are given in the Instructor's Guide for Electrical Motor Controls Manual.

IDENTIFICATION
Select the response that matches the given word, symbol, etc. Write the response in the space provided.

List the reference wire number for each wire on the line diagram and pushbutton station. Mark each wire except the wire connecting the starting coil to the overload contacts.

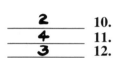

2 10.
4 11.
3 12.

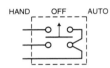

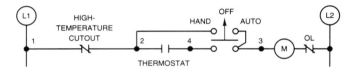

COMPLETION
Select the response that completes the statement. Write the answer in the space provided.

Answer the questions using Models E and F Single Shot Relays on page 60.

2 and 7   1. What are the pin numbers of the timer coil?

10   2. The maximum current that can be switched with a Model E timer is _____ A.

CALCULATIONS
Solve the problem(s) based on the information given. Write the answer in the space provided.

Solve the problems using Ohm's Law or the Power Formula.

125   7. E = _____ V

R = 25 kΩ

I = 5 mA

12   8. P = _____ mW

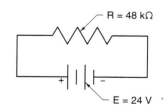

R = 48 kΩ

E = 24 V

SHORT ANSWER

Select the response that completes the statement. Write the answer in the space provided.

Identify the roller that must be adjusted for proper belt tracking. Show the direction of adjustment on the drawing.

head end snub roller **1.** Roller to be adjusted.   head end snub roller **2.** Roller to be adjusted.

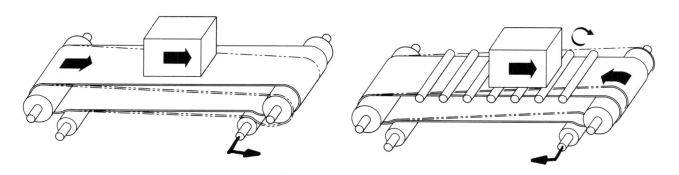

ILLUSTRATED ANSWER

Complete the problem(s) by adding the appropriate lines, letters, numbers, etc.

2. Connect the transformer for 240 V to 120 V.   **3.** Connect the transformer for 240 V to 240 V.

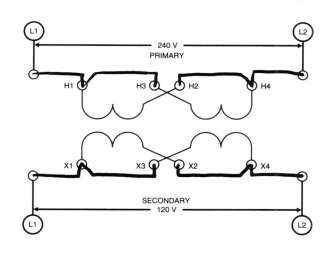

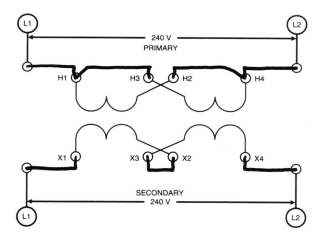

Application — Electrical Prefixes

Prefixes

Measured or calculated electrical units may be large or small. For example, solid-state devices may have a current draw of less than 0.000001 amperes (A). In an industrial plant that melts aluminum, power greater than 100,000 watts (W) may be used.

To avoid long expressions, prefixes are used to indicate units that are smaller and larger than the base unit. For example, 0.000001 A equals 1 microampere (μA), and 100,000 W is equal to 100 kilowatts (kW). **See Common Prefixes.**

Converting Units

To convert between different units, move the decimal point to the left or right depending on the unit. **See Conversion Table.**

Common Electrical Quantities

Abbreviations are used for common electrical quantities to simplify their expression. **See Common Electrical Quantities.**

COMMON PREFIXES		
Symbol	Prefix	Equivalent
G	giga	1,000,000,000
M	mega	1,000,000
k	kilo	1000
base unit	—	1
m	milli	0.001
μ	micro	0.000001
n	nano	0.000000001
p	pico	0.000000000001

Example: Converting Units

Convert .000001 A to simplest terms.

Move the decimal point six places to the right to obtain 1.0 μA (from Conversion Table).

.000001 A = **1.0 μA** or **1 μA**

Example: Electrical Abbreviations

Abbreviate the following electrical terms.

120 milliwatts = **120 mW**

120 watts = **120 W**

50 farads = **50 F**

120 kilovolts = **120 kV**

10 amperes = **10 A**

CONVERSION TABLE								
Initial Units	Final Units							
	giga	mega	kilo	base unit	milli	micro	nano	pico
giga		3R	6R	9R	12R	15R	18R	21R
mega	3L		3R	6R	9R	12R	15R	18R
kilo	6L	3L		3R	6R	9R	12R	15R
base unit	9L	6L	3L		3R	6R	9R	12R
milli	12L	9L	6L	3L		3R	6R	9R
micro	15L	12L	9L	6L	3L		3R	6R
nano	18L	15L	12L	9L	6L	3L		3R
pico	21L	18L	15L	12L	9L	6L	3L	

R = Move the decimal point to the right.
L = Move the decimal point to the left.

COMMON ELECTRICAL QUANTITIES		
Variable	Name	Unit of Measure and Abbreviation
E	voltage	volts — E
I	current	amperes — A
R	resistance	ohms — Ω
P	power	watts — W
P	power (apparent)	volt-amps — VA
C	capacitance	farads — F
L	inductance	henrys — H
Z	impedance	ohms — Ω

Application — Using Ohm's Law and the Power Formula

Ohm's Law

Ohm's law is the relationship between the voltage, current, and resistance in an electrical circuit. Ohm's law states that current in a circuit is proportional to the voltage and inversely proportional to the resistance. It is written $I = \dfrac{E}{R}$, $R = \dfrac{E}{I}$, and $E = R \times I$.

Power Formula

The *power formula* is the relationship between the voltage, current, and power in an electrical circuit. The power formula states that the power in a circuit is equal to the voltage times the current. It is written $P = E \times I$, $E = \dfrac{P}{I}$, and $I = \dfrac{P}{E}$.

Any value in these relationships is found using Ohm's Law and Power Formula. **See Ohm's Law and Power Formula.**

P = WATTS
I = AMPS
R = OHMS
E = VOLTS

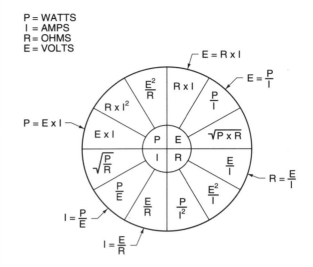

VALUES IN INNER CIRCLE
ARE EQUAL TO VALUES IN
CORRESPONDING OUTER CIRCLE

OHM'S LAW AND POWER FORMULA

Example: Finding Current using Ohm's Law

A circuit has a voltage supply of 120 V and a resistance of 60 Ω. Find the current in the circuit.

$$I = \frac{E}{R}$$

$$I = \frac{120}{60}$$

$$I = \textbf{2 A}$$

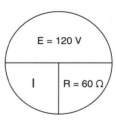

Example: Finding Power using Power Formula

A load that draws 5 A is connected to a 240 V power supply. Find the power in the circuit.

$$P = E \times I$$

$$P = 240 \times 5$$

$$P = \textbf{1200 W}$$

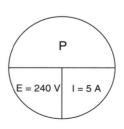

◰ Application — Multimeter Use

Multimeter

Electrical measurements are required when installing, operating, or repairing electrical equipment. A multimeter is the most common meter used to take electrical measurements. They are lightweight, portable, and can be used to measure alternating and direct current. A *multimeter* is a device that measures voltage (E), current (A), and resistance (Ω). The function and range switches must be set at the correct quantity, and the correct scale must be read when taking electrical measurements using a multimeter. **See Multimeter.**

Electrical Quantities Measured

The three electrical quantities measured with most multimeters are voltage, current, and resistance. The function switch is set on AC when alternating voltage (VAC) is measured, and is set on DC when direct voltage, or current, (DC) is measured. If the multimeter does not have a separate setting for resistance, the positive DC setting is used. **See Function Switch.**

The normal setting for measuring DC voltage or current is +DC. This setting makes the red (+ marked) lead positive.

The alternative setting for measuring DC voltage or current is −DC. This setting makes the red (+ marked) lead negative. The setting is used when it is preferable to have the test lead with the alligator clip (black) positive and the test lead with the pointer (red) negative.

Scales

Most multimeters have separate scales for reading voltage, current, and resistance. The scales are usually labelled on both ends. Once the correct scale is determined, the number of units represented by the scale divisions must be determined. **See Scales.**

For example, there are 10 divisions between 0 and 10 on the Ω scale. The value of each division is found by dividing 10 by 10 which equals 1 Ω per division (10 ÷ 10 = 1). This value is then multiplied by the range switch setting to obtain the correct value.

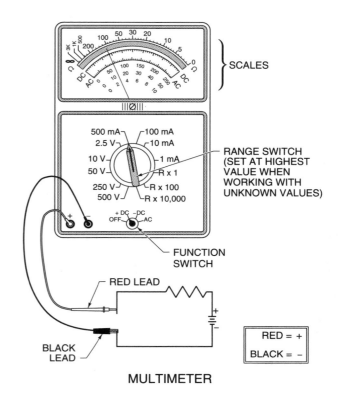

MULTIMETER

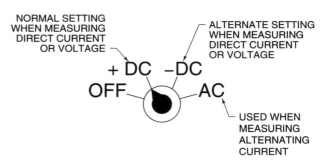

FUNCTION SWITCH

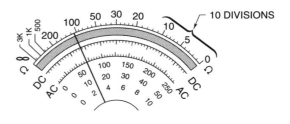

SCALES

Application — Resistance in a Series Circuit

Series Circuit

A *series circuit* is a circuit that contains two or more loads and one path through which current flows. The total resistance of the loads (resistors) connected in series is equal to the sum of the individual resistances. The total resistance of a series circuit is found by applying the formula:

$$R_T = R_1 + R_2 + R_3 + \ldots$$

where

R_T = total resistance (in ohms)

R_1 = resistance 1 (in ohms)

R_2 = resistance 2 (in ohms)

R_3 = resistance 3 (in ohms)

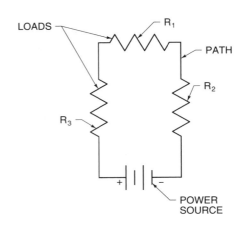

Example: Finding Total Resistance — Series Circuit

A circuit has four resistors of 10 Ω, 55 Ω, 100 Ω, and 800 Ω connected in series. Find the total resistance in the circuit.

$$R_T = R_1 + R_2 + R_3 + R_4$$
$$R_T = 10 + 55 + 100 + 800$$
$$R_T = \mathbf{965\ \Omega}$$

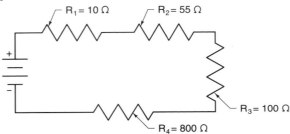

Application — Resistance in a Parallel Circuit

Parallel Circuit

A *parallel circuit* is a circuit that contains two or more loads and has more than one path through which current flows. The total resistance of a parallel circuit with two resistors is found by applying the formula:

$$R_T = \frac{R_1 \times R_2}{R_1 + R_2}$$

The total resistance of a parallel circuit with three or more resistors is found by applying the formula:

$$R_T = \frac{1}{\dfrac{1}{R_1} + \dfrac{1}{R_2} + \dfrac{1}{R_3} + \ldots}$$

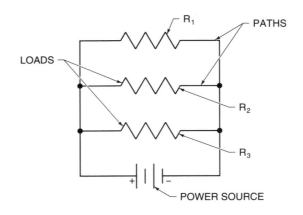

Example: Finding Total Resistance — Parallel Circuit

A circuit has two resistors of 50 Ω and 200 Ω connected in parallel. Find the total resistance in the circuit.

$$R_T = \frac{R_1 \times R_2}{R_1 + R_2} \qquad\qquad R_T = \frac{10,000}{250}$$

$$R_T = \frac{50 \times 200}{50 + 200} \qquad\qquad R_T = \mathbf{40\ \Omega}$$

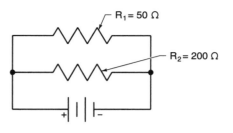

Application — Resistance in a Series/Parallel Circuit

Series/Parallel Circuit

Most electrical circuits are combinations of series and parallel circuits. The total resistance of a series/parallel circuit is found by calculating the equivalent resistance of the parallel circuit(s) and add the value to the resistance of the loads connected in series. The total resistance in a series/parallel circuit is found by applying the formula:

$$R_T = \frac{R_{P1} \times R_{P2}}{R_{P1} + R_{P2}} + R_{S1} + R_{S2} + \dots$$

where

R_T = total resistance (in ohms)

R_{P1} = parallel resistance 1 (in ohms)

R_{P2} = parallel resistance 2 (in ohms)

R_{S1} = series resistance 1 (in ohms)

R_{S2} = series resistance 2 (in ohms)

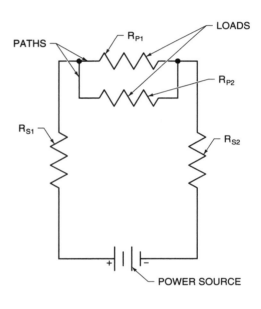

Example: Finding Total Resistance — Series/Parallel Circuit

A circuit has two resistors of 150 Ω and 300 Ω connected in parallel and three resistors of 75 Ω, 50 Ω, and 25 Ω connected in series. Find the total resistance in the circuit.

$$R_T = \frac{R_{P1} \times R_{P2}}{R_{P1} + R_{P2}} + R_{S1} + R_{S2} + R_{S3}$$

$$R_T = \frac{150 \times 300}{150 + 300} + 75 + 50 + 25$$

$$R_T = \frac{45,000}{450} + 150$$

$$R_T = 100 + 150$$

$$R_T = \mathbf{250\ \Omega}$$

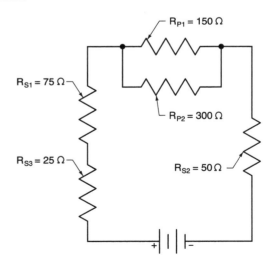

Application — Temperature Conversion

Converting Fahrenheit to Celsius

Converting between Fahrenheit and Celsius temperatures is often required in calculations and when using manufacturer's data. To convert a Fahrenheit temperature reading to Celsius, subtract 32 from the Fahrenheit reading and divide by 1.8. To convert Fahrenheit to Celsius, apply the formula:

$$^\circ C = \frac{^\circ F - 32}{1.8}$$

where

$^\circ C$ = degrees Celsius

$^\circ F$ = degrees Fahrenheit

32 = difference between bases

1.8 = ratio between bases

Example: Converting Fahrenheit to Celsius

TW insulated conductor is rated at 140°F. Convert the Fahrenheit temperature to Celsius.

$$^\circ C = \frac{^\circ F - 32}{1.8}$$

$$^\circ C = \frac{140 - 32}{1.8}$$

$$^\circ C = \frac{108}{1.8}$$

$$^\circ C = \mathbf{60^\circ C}$$

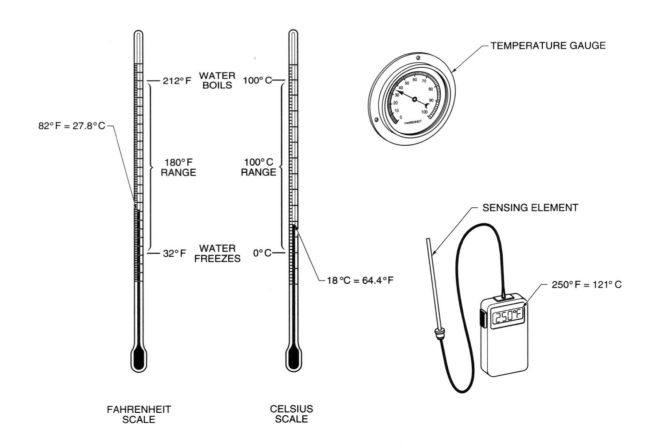

Converting Celsuis to Fahrenheit

To convert a Celsius temperature reading to Fahrenheit, multiply 1.8 by the Celsius reading and add 32. To convert Celsius to Fahrenheit, apply the formula:

$$°F = (1.8 \times °C) + 32$$

where

$°F$ = degrees Fahrenheit

1.8 = ratio between bases

$°C$ = degrees Celsius

32 = difference between bases

Example: Converting Celsius to Fahrenheit

A motor has a permissible temperature rise of 40°C. Convert the Celsius temperature to Fahrenheit.

$$°F = (1.8 \times °C) + 32$$
$$°F = (1.8 \times 40) + 32$$
$$°F = 72 + 32$$
$$°F = \mathbf{104°F}$$

▢ Application — Reading Graphs

Graphs

A *graph* is a diagram that shows the continuous relationship between two or more variables. Graphs present information in a simple form and are commonly used by component manufacturers to illustrate data and specifications. On a graph, one known variable is plotted horizontally and another is plotted vertically. The relationship between the two variables is represented by a straight line or curved line. The point at which either variable line intersects the straight or curved line represents the value of the unknown variable.

A graph may be used to illustrate the effect of the ambient temperature on the operating characteristics of a fuse. *Ambient temperature* is the temperature of the surrounding air in which the fuse is installed. As the ambient temperature increases, the opening time and capacity rating of the fuse decreases. As the ambient temperature decreases, the opening time and capacity rating of the fuse increases. For example, at 140°F, the opening time and capacity rating of a fuse is 70% of the standard rated opening time. Likewise, at −40°F, the opening time and capacity rating of the fuse is 130% of the standard rated opening time. **See Fuse Graph.**

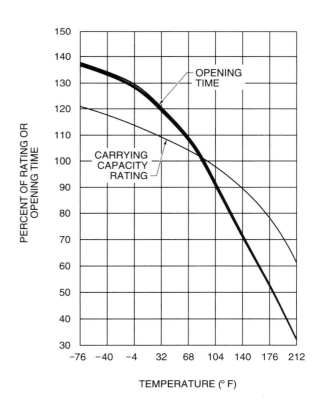

FUSE GRAPH

Plotting Graphs

Relationships between variables are often given in tables. Tables present the relationship between several variables. **See Table.** To find the continuous relationship between the variables, convert the table to a graph. A table is converted to a graph by applying the procedure:

1. Draw two axes at right angles.
2. Label the two axes.
3. Select appropriate scales based on the values given.
4. Plot the points.
5. Draw a curve through the points.

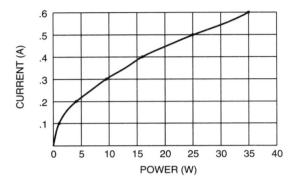

CURRENT (A)	.1	.2	.3	.4	.5	.6
POWER (W)	1	4	9	11	25	36

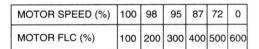

Example: Plotting a Graph

The starting current draw of a motor varies with the speed of the motor. Plot the graph of the relationship.

1. Draw two axes at right angles. **See Graph.**
2. Label the two axes.

   Label the vertical axis motor speed and the horizontal axis motor full load current.

3. Select appropriate scales based on the values given.

   Select 0 to 100 for the motor speed scale and 0 to 600 for the motor full load current scale.

4. Plot the points.
5. Draw a curve through the points.

MOTOR SPEED (%)	100	98	95	87	72	0
MOTOR FLC (%)	100	200	300	400	500	600

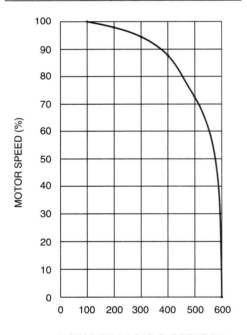

ACTIVITIES

Name _____   Date _____

⬤ Activity 1-1.   Electrical Prefixes

State each value in its simplest form.

_____ **1.** 0.045 V = _____

_____ **2.** 22,000 Ω = _____

_____ **3.** 0.006 A = _____

_____ **4.** 0.0004 V = _____

_____ **5.** 0.00000052 A = _____

_____ **6.** 21,000,000,000 W = _____

_____ **7.** 0.00000000004 V = _____

_____ **8.** 0.005 V = _____

_____ **9.** 2000 V = _____

_____ **10.** 0.00000003 A = _____

_____ **11.** 0.00360 A = _____

_____ **12.** 3,300,000 Ω = _____

_____ **13.** 0.00005 W = _____

_____ **14.** 0.600000 V = _____

_____ **15.** 0.000000002 W = _____

_____ **16.** 35,100,000 W = _____

_____ **17.** 3005 V = _____

_____ **18.** 1000.5 A = _____

_____ **19.** 10.050 A = _____

_____ **20.** 0.1001 W = _____

⬤ Activity 1-2.   Using Ohm's Law and the Power Formula

Solve the problems using Ohm's Law or the Power Formula.

_____ **1.** E = _____ V

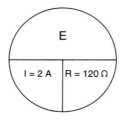

_____ **2.** I = _____ mA

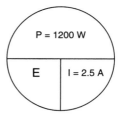

_____ **3.** P = _____ mW

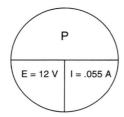

_____ **4.** E = _____ V

_____   **5.** I = _____ mA

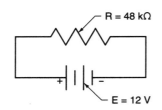

R = 48 kΩ

\+ | | | −

E = 12 V

_____   **6.** R = _____ Ω

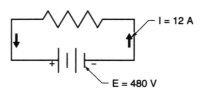
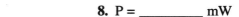

I = 12 A

\+ | | | −

E = 480 V

_____   **7.** E = _____ V

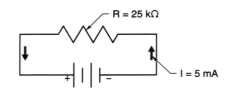

R = 25 kΩ

\+ | | | −

I = 5 mA

_____   **8.** P = _____ mW

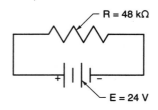

R = 48 kΩ

\+ | | | −

E = 24 V

_____   **9.** E = _____ V
        I = 50 mA
        R = 10 kΩ

_____   **10.** E = 26 V
        I = 60 mA
        R = _____ Ω

_____   **11.** E = 100 V
        I = _____ A
        P = 100 mW

_____   **12.** E = 12 V
        I = 50 mA
        P = _____ mW

_____   **13.** E = 24 V
        I = _____ A
        P = 24 mW

_____   **14.** E = _____ V
        I = 3 A
        P = 45 W

_____   **15.** E = 120 V
        I = 100 mA
        P = _____ W

_____   **16.** E = 480 V
        I = _____ A
        P = 1200 W

_____   **17.** E = 220 V
        I = 4000 mA
        R = _____ Ω

_____   **18.** E = _____ mV
        I = 15 µA
        R = 15 kΩ

_____   **19.** E = _____ V
        I = 7 µA
        R = 1 MΩ

_____   **20.** E = _____ kV
        I = 2 mA
        R = 5 MΩ

_____   **21.** E = 480 V
        I = _____ A
        P = 12 kW

_____   **22.** E = 240 V
        I = 125 A
        P = _____ MW

_____   **23.** E = 36 V
        I = _____ A
        P = 198 W

_____   **24.** E = _____ V
        I = 250 A
        P = 50 kW

_____   **25.** E = 500 mV
        I = 500 mA
        P = _____ W

_____   **26.** E = 208 V
        I = _____ A
        P = 6240 W

⬤ **Activity 1-3.   Multimeter Use**

List the correct reading for the function/range settings.

_____ **1.** Reading = _____ mA (DC)

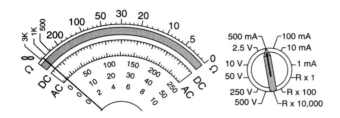

_____ **2.** Reading = _____ μ A (DC)

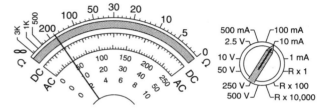

_____ **3.** Reading = _____ kΩ

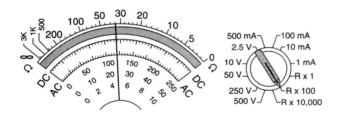

_____ **4.** Reading = _____ VAC

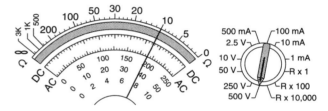

_____ **5.** Reading = _____ VDC

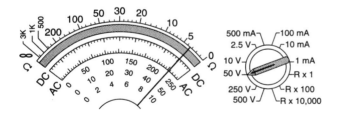

_____ **6.** Reading = _____ mA (DC)

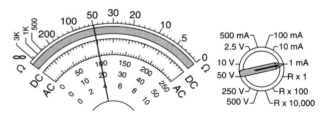

_____ **7.** Reading = _____ VDC

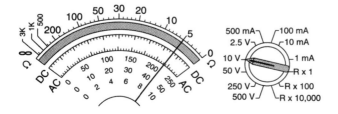

_____ **8.** Reading = _____ mA (DC)

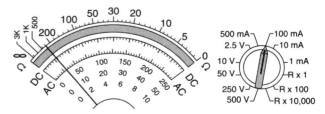

Activity 1-4. Resistance in a Series Circuit

Determine the total resistance in Series Circuit 1.

_____ **1.** $R_T =$ _____ Ω

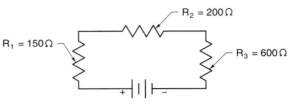

SERIES CIRCUIT 1

Determine the total resistance and unknown electrical quantity in Series Circuit 2.

_____ **2.** $R_T =$ _____ Ω
_____ **3.** $I_T =$ _____ mA

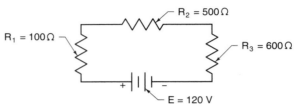

SERIES CIRCUIT 2

Determine the total resistance in Series Circuit 3.

_____ **4.** $R_T =$ _____ Ω

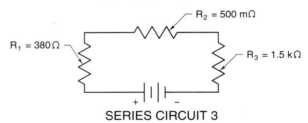

SERIES CIRCUIT 3

Determine the total resistance and unknown electrical quantities in Series Circuit 4.

_____ **5.** $R_T =$ _____ kΩ
_____ **6.** $I_T =$ _____ mA
_____ **7.** $P_T =$ _____ W

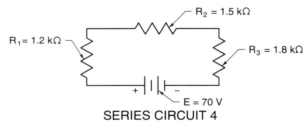

SERIES CIRCUIT 4

 Activity 1-5. Resistance in a Parallel Circuit

Determine the total resistance in Parallel Circuit 1.

_____ **1.** $R_T =$ _____ Ω

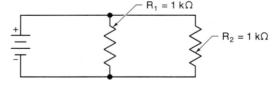

PARALLEL CIRCUIT 1

Determine the total resistance and unknown electrical quantity in Parallel Circuit 2.

_____ **2.** $R_T =$ _____ Ω
_____ **3.** $I_T =$ _____ mA

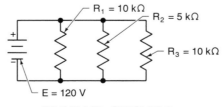

PARALLEL CIRCUIT 2

Determine the total resistance in Parallel Circuit 3.

_____ **4.** $R_T =$ _____ Ω

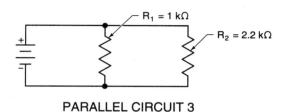

PARALLEL CIRCUIT 3

Determine the total resistance and unknown electrical quantity in Parallel Circuit 4.

_____ **5.** $R_T =$ _____ Ω
_____ **6.** $I_T =$ _____ mA
_____ **7.** $P_T =$ _____ W

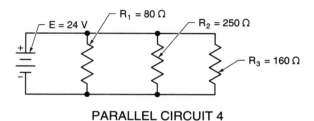

PARALLEL CIRCUIT 4

⬤ Activity 1-6. Resistance in a Series/Parallel Circuit

Determine the total resistance in Series/Parallel Circuit 1.

_____ **1.** $R_T =$ _____ Ω

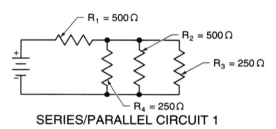

SERIES/PARALLEL CIRCUIT 1

Determine the total resistance in Series/Parallel Circuit 2.

_____ **2.** $R_T =$ _____ Ω

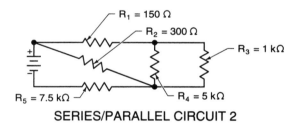

SERIES/PARALLEL CIRCUIT 2

⬤ Activity 1-7. Temperature Conversion

_____ **1.** $20°C =$ _____ $°F$
_____ **2.** An ambient temperature of $40°C$ is _____ $°F$.
_____ **3.** $460°C =$ _____ $°F$
_____ **4.** $140°F =$ _____ $°C$
_____ **5.** No. 8 THHN copper conductor can be used in temperatures below _____ $°F$ ($186°C$).
_____ **6.** A temperature of $77°F$ is _____ $°C$.
_____ **7.** $1998°F =$ _____ $°C$
_____ **8.** The maximum operating temperature of FFH-2 fixture wire is _____ $°C$ ($167°F$).
_____ **9.** $1610°C =$ _____ $°F$
_____ **10.** A temperature of $32°F$ is _____ $°C$.

◑ Activity 1-8. Reading Graphs

Complete the statements and answer the questions using Fuse Graph on page 7.

_____ **1.** If a fuse is installed in an ambient temperature of 140°F, its current-carrying capacity is decreased by _____%.

_____ **2.** If a fuse is installed in an ambient temperature of 140°F, its opening time is decreased by _____%.

_____ **3.** If a 10 A rated fuse is installed in an ambient temperature of 32°F, the fuse carries a(n) _____ A load before opening.

_____ **4.** Would a 10 A rated fuse installed in an ambient temperature of 32°F respond faster or slower than its rated opening time to an overcurrent?

_____ **5.** If a 10 A rated fuse is installed in an ambient temperature of 176°F, the fuse carries a(n) _____ A load before opening.

_____ **6.** Would a 10 A fuse installed in an ambient temperature of 176°F respond faster or slower than its rated opening time to an overcurrent?

◐ Activity 1-9. Plotting Graphs

Complete the tables and plot the relationships on the graphs.

1.

R =(Ω)	10	10	10	10	10	10	10	10	10	10
E =(V)	10	20	30	40	50	60	70	80	90	100
I =(A)										

2.

R =(Ω)	100	100	100	100	100	100	100	100	100	100
E =(V)	2	4	6	8	10	12	14	16	18	20
I =(A)										

CURRENT (A)

VOLTAGE (V)

CURRENT (A)

VOLTAGE (V)

APPLICATIONS

▢ Application — Resistor Color Coding

Resistors are devices that limit the current flowing in an electronic circuit. Small resistors use color bands to represent their resistance value. The first two color bands represent the first two digits in the value of the resistor. The third color band (multiplier) indicates the number of zeros that must be added to the first two digits. The fourth band (tolerance) indicates how far the actual measured value can be from the coded value. **See Resistor Color Codes.**

	RESISTOR COLOR CODES			
COLOR	1st NUMBER	2nd NUMBER	MULTIPLIER	TOLER-ANCE (%)
Black	0	0	1	0
Brown	1	1	10	
Red	2	2	100	
Orange	3	3	1000	
Yellow	4	4	10,000	
Green	5	5	100,000	
Blue	6	6	1,000,000	
Violet	7	7	10,000,000	
Gray	8	8	100,000,000	
White	9	9	1,000,000,000	
Gold			0.1	5
Silver			0.01	10
None			0	20

Example: Finding Resistor Value

A resistor with a red, black, orange, and silver color band has a resistance value of 20,000 Ω (20 kΩ) ±10%.

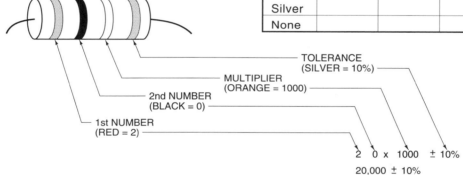

TOLERANCE (SILVER = 10%)
MULTIPLIER (ORANGE = 1000)
2nd NUMBER (BLACK = 0)
1st NUMBER (RED = 2)

2   0   x   1000   ± 10%

20,000 ± 10%

▢ Application — Terminal Block Wire Connections

Electrical connections are made using solder connections, clamp springs, and terminal blocks. A terminal block provides a convenient place to make wire changes and perform tests when troubleshooting. **See Wiring Diagram.**

Terminal Block Tables

Manufacturers provide tables that determine the maximum number of wires that can be connected to a terminal block. **See Terminal Blocks.**

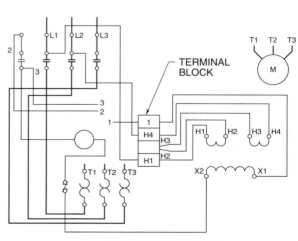

WIRING DIAGRAM

15

					TERMINAL BLOCKS							
	Wire Size (AWG)											
Catalog Number	#22	#20	#18	#16	#14	#12	#10	#8	#6	#4	#2	#1/0
	Number of the same size wires per terminal block											
001	3	3	3	2	1	—	—	—	—	—	—	—
002	4	3	3	3	2	1	—	—	—	—	—	—
003	4	4	4	3	2	2	1	1	—	—	—	—
004	—	4	4	3	2	2	1	1	—	—	—	—
005	5	5	4	4	3	2	2	1	1	—	—	—
006	—	—	—	—	—	4	4	4	3	2	1	1
007	4	4	4	3	2	2	1	—	—	—	—	—
008	4	4	3	2	2	1	—	—	—	—	—	—
009	4	4	4	3	3	2	1	—	—	—	—	—
010	6	5	5	4	3	2	1	—	—	—	—	—

▢ Application — Symbol and Abbreviation Identification

Electrical prints are used when designing, troubleshooting, servicing, or repairing circuits. These prints use standard symbols and abbreviations to show circuit operation and device use. **See Selected Symbols and Abbreviations. See Electrical Symbols and Abbreviations in Appendix.**

						SELECTED SYMBOLS

SELECTED SYMBOLS table with columns: LIMIT SWITCHES (NO, NC — HELD CLOSED, HELD OPEN), PRESSURE AND VACUUM SWITCHES, TEMPERATURE-ACTIVATED SWITCH, TIMED CONTACTS ENERGIZED (NOTC, NCTO), THERMAL OVERLOAD RELAY, CONTROL TRANSFORMER SINGLE VOLTAGE (H1 H2 / X2 X1)

SELECTED ABBREVIATIONS

AC	ALTERNATING CURRENT	NO	NORMALLY OPEN
CB	CIRCUIT BREAKER	OL	OVERLOAD RELAY
CR	CONTROL RELAY	PB	PUSHBUTTON
DC	DIRECT CURRENT	PS	PRESSURE SWITCH
DP	DOUBLE POLE	R	REVERSE
DPST	DOUBLE POLE, SINGLE THROW	S	SWITCH
F	FORWARD	SOL	SOLENOID
LS	LIMIT SWITCH	SP	SINGLE POLE
M	MOTOR STARTER	SPDT	SINGLE POLE, DOUBLE THROW
MTR	MOTOR	SPST	SINGLE POLE, SINGLE THROW
NC	NORMALLY CLOSED	TR	TIME DELAY RELAY

ACTIVITIES

Name _____ Date _____

2

○ Activity 2-1. Resistor Color Coding

State the resistance, tolerance, and resistance in simplest form using Resistor Color Codes on page 15.

1. red, red, yellow, gold

_____ **A.** Resistance is _____ Ω.

_____ **B.** Tolerance is _____ %.

_____ **C.** The simplest form is _____.

2. blue, green, green, gold

_____ **A.** Resistance is _____ Ω.

_____ **B.** Tolerance is _____ %.

_____ **C.** The simplest form is _____.

3. brown, black, red, silver

_____ **A.** Resistance is _____ Ω.

_____ **B.** Tolerance is _____ %.

_____ **C.** The simplest form is _____.

4. black, brown, brown, gold

_____ **A.** Resistance is _____ Ω.

_____ **B.** Tolerance is _____ %.

_____ **C.** The simplest form is _____.

5. white, brown, gray, gold

_____ **A.** Resistance is _____ Ω.

_____ **B.** Tolerance is _____ %.

_____ **C.** The simplest form is _____.

6. violet, green, red, gold

_____ **A.** Resistance is _____ Ω.

_____ **B.** Tolerance is _____ %.

_____ **C.** The simplest form is _____.

7. yellow, blue, brown, gold

_____ **A.** Resistance is _____ Ω.

_____ **B.** Tolerance is _____ %.

_____ **C.** The simplest form is _____.

8. orange, gray, yellow, gold

_____ **A.** Resistance is _____ Ω.

_____ **B.** Tolerance is _____ %.

_____ **C.** The simplest form is _____.

9. gray, white, black, silver

_____ **A.** Resistance is _____ Ω.

_____ **B.** Tolerance is _____ %.

_____ **C.** The simplest form is _____.

10. green, red, orange, silver

_____ **A.** Resistance is _____ Ω.

_____ **B.** Tolerance is _____ %.

_____ **C.** The simplest form is _____.

○ Activity 2-2. Terminal Block Wire Connections

Answer the questions using Terminal Blocks on page 16.

_____ **1.** The terminal block that can connect three #6 wires is number _____.

_____ **2.** The terminal block that can connect six #22 wires is number _____.

_____ **3.** The terminal block that can connect five #18 wires is number _____.

_____ **4.** The maximum number of #14 wires that can be connected using a #001 terminal block is _____.

_____ **5.** The maximum number of #14 wires that can be connected using a #007 terminal block is _____.

_____ **6.** Can a pair of #2 wires be connected using a #006 terminal block?

○ Activity 2-3. Symbol and Abbreviation Identification

Identify the symbols using Starter Wiring Diagram.

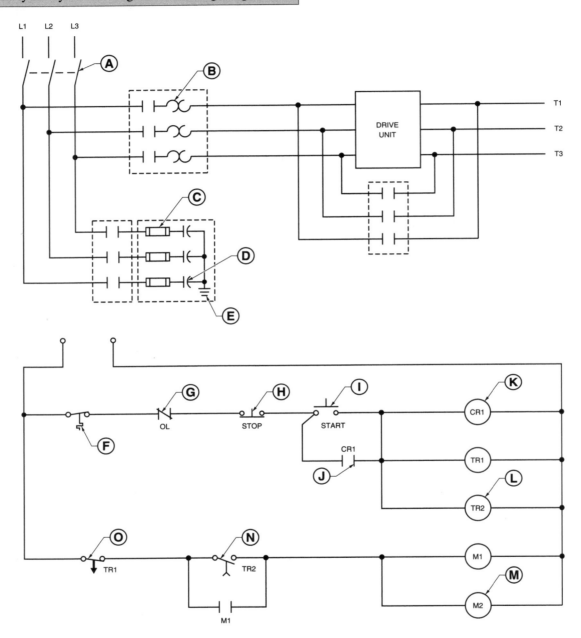

STARTER WIRING DIAGRAM

_____ 1. Normally open timed contacts

_____ 2. Normally closed timed contacts

_____ 3. Normally closed temperature actuated switch

_____ 4. Normally open pushbutton

_____ **6.** Normally open relay contacts

_____ **7.** Overload contact

_____ **8.** Motor starter

_____ **9.** Capacitor

_____ **10.** Ground

_____ **11.** Relay coil

_____ **12.** Timer coil

_____ **13.** Fuse

_____ **14.** Thermal overload

_____ **15.** Disconnect

Identify the abbreviations using Solid-state Wiring Diagram.

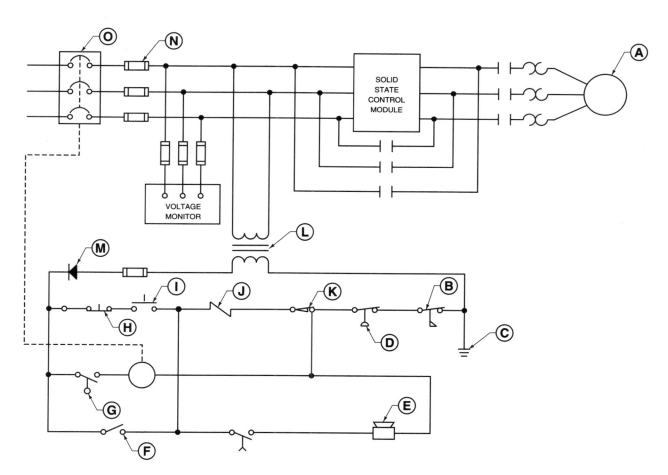

SOLID-STATE WIRING DIAGRAM

_____   **16.** T
_____   **17.** CB
_____   **18.** FU
_____   **19.** PB-NO
_____   **20.** PB-NC
_____   **21.** FS
_____   **22.** LS
_____   **23.** ALM

_____   **24.** GRD
_____   **25.** SOL
_____   **26.** SPST
_____   **27.** FLS
_____   **28.** PS
_____   **29.** DIO
_____   **30.** MTR

31. Add the electrical symbols to Control Wiring Diagram.

A. Normally open foot switch
B. Pilot light
C. Normally open flow switch
D. Normally closed vacuum switch
E. Control fuse
F. Normally closed limit switch

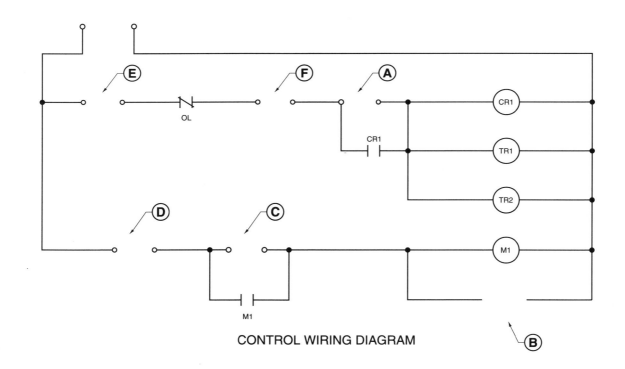

CONTROL WIRING DIAGRAM

Application — Assigning Wire Reference Numbers

Line Diagrams

A *line diagram* (ladder diagram) is a diagram that shows the logic of a control circuit in simplest form. A line diagram does not show the location of each component in relationship to the other components in the circuit. A line diagram is used when designing, modifying, or explaining a circuit.

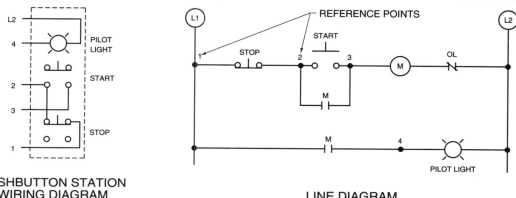

PUSHBUTTON STATION
WIRING DIAGRAM

LINE DIAGRAM

Assigning Reference Numbers

Each wire in a control circuit is assigned a reference point on a line diagram to keep track of the different wires that connect the components in the circuit. Each reference point is assigned a reference number. Reference numbers are normally assigned from the top left to the bottom right.

When assigning wire reference numbers, any wire that is always connected to a point is assigned the same number. The wires that are assigned a number vary from 2 to the number required by the circuit. Any wire that is prewired when the component is purchased is normally not assigned a reference number. When assigning reference numbers, different numbering assignments can be used. The exact numbering system varies for each manufacturer or design engineer. This numbering system applies to any control circuit such as single station, multistation, or reversing circuits.

For example, a line diagram has five reference points. The reference points are assigned the numbers 1, 2, 3, 4, and L2. The first reference point is labeled 1 or L1. Any wire connected to this point at all times is labeled 1. The five reference points on the circuit could be given any of the following number sets. **See Line Diagrams.**

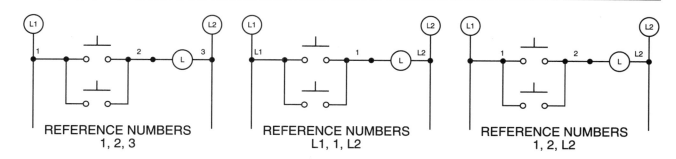

REFERENCE NUMBERS
1, 2, 3

REFERENCE NUMBERS
L1, 1, L2

REFERENCE NUMBERS
1, 2, L2

Application — Basic Switching Logic

Logic Functions

Control circuits are designed to perform a specific function. *Logic* is the way a circuit functions. Common logic functions are applied to different electrical circuits. Names for common logic functions include AND, OR, NOT, NOR, and NAND. The logic function depends on the relationship between the input and output signals of a circuit.

Inputs are the switches that start or stop the flow of electricity to the outputs. *Outputs* are the loads that use the electricity delivered by the switches to produce work. Typical loads are lights, motors, heating elements, and solenoids. A circuit is activated when the switch contacts are switched manually (pushbutton), mechanically (limit switch), or automatically (temperature switch). **See Basic Logic Functions and Appendix.**

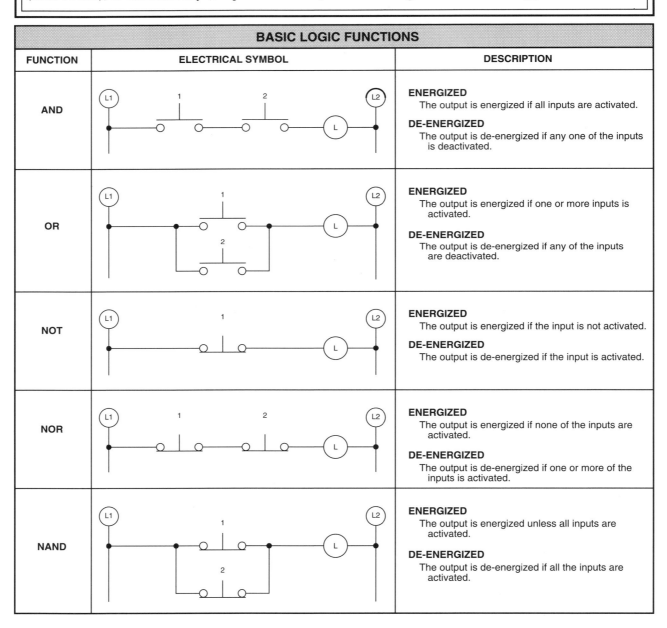

BASIC LOGIC FUNCTIONS		
FUNCTION	**ELECTRICAL SYMBOL**	**DESCRIPTION**
AND		**ENERGIZED** The output is energized if all inputs are activated. **DE-ENERGIZED** The output is de-energized if any one of the inputs is deactivated.
OR		**ENERGIZED** The output is energized if one or more inputs is activated. **DE-ENERGIZED** The output is de-energized if any of the inputs are deactivated.
NOT		**ENERGIZED** The output is energized if the input is not activated. **DE-ENERGIZED** The output is de-energized if the input is activated.
NOR		**ENERGIZED** The output is energized if none of the inputs are activated. **DE-ENERGIZED** The output is de-energized if one or more of the inputs is activated.
NAND		**ENERGIZED** The output is energized unless all inputs are activated. **DE-ENERGIZED** The output is de-energized if all the inputs are activated.

ACTIVITIES

Name _____ Date _____

○ Activity 3-1. Assigning Wire Reference Numbers

List the reference number for each wire coming from the pushbutton station.

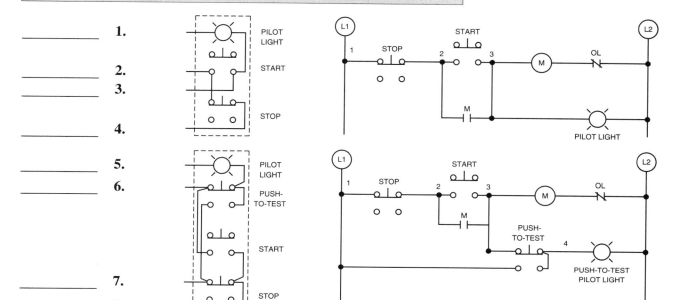

_____ 1.

_____ 2.

_____ 3.

_____ 4.

_____ 5.

_____ 6.

_____ 7.

_____ 8.

○ Activity 3-2. Numbering Control Circuits — Single Station

List the reference wire number for each wire on the line diagram and pushbutton station. Mark each wire except the wire connecting the starting coil to the overload contacts.

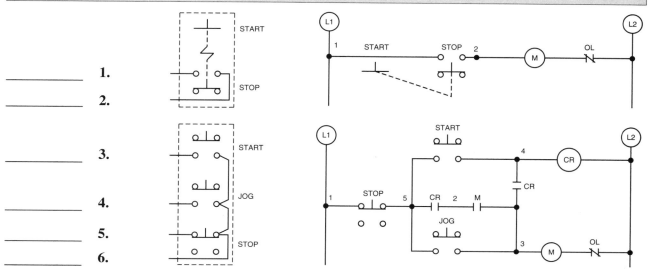

_____ 1.

_____ 2.

_____ 3.

_____ 4.

_____ 5.

_____ 6.

_____ **7.**

_____ **8.**

_____ **9.**

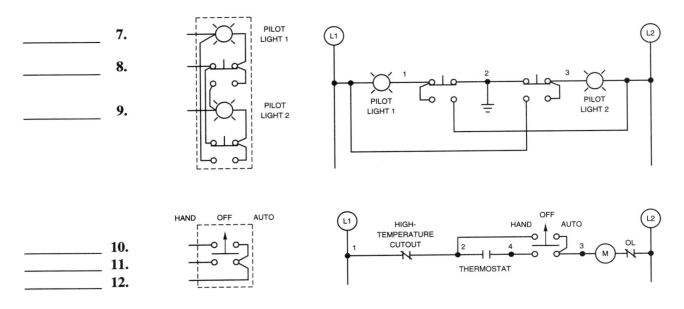

_____ **10.**
_____ **11.**
_____ **12.**

⬤ Activity 3-3.   Numbering Control Circuits — Multistation

List the reference wire number for each wire on the line diagram and pushbutton station. Mark each wire except the wire connecting the starting coil to the overload contacts.

_____ **1.**

_____ **2.**

_____ **3.**

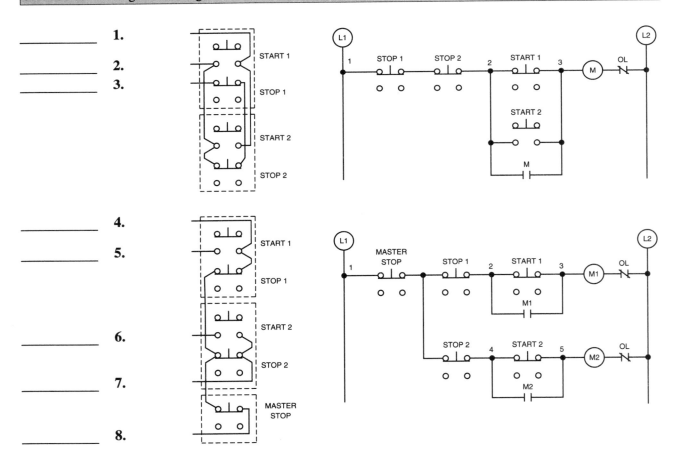

_____ **4.**

_____ **5.**

_____ **6.**

_____ **7.**

_____ **8.**

◯ Activity 3-4. Numbering Control Circuits — Reversing

List the reference wire number for each wire on the line diagram and pushbutton station. Mark each wire except the wire connecting the starting coil to the overload contacts.

_____ 1.

_____ 2.

_____ 3.

_____ 4.

_____ 5.

_____ 6.

_____ 7.

_____ 8.

_____ 9.

_____ 10.

_____ 11.

_____ 12.

_____ 13.

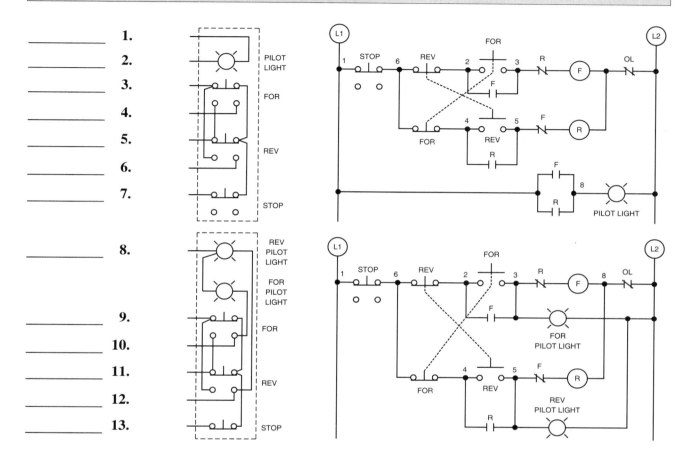

◯ Activity 3-5. Numbering Control Circuits — Two-speed

List the reference wire number for each wire on the line diagram and pushbutton station. Mark each wire except the wire connecting the starting coil to the overload contacts.

_____ 1.

_____ 2.

_____ 3.

_____ 4.

_____ 5.

_____ 6.

_____ 7.

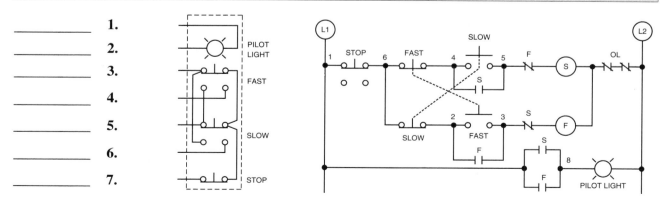

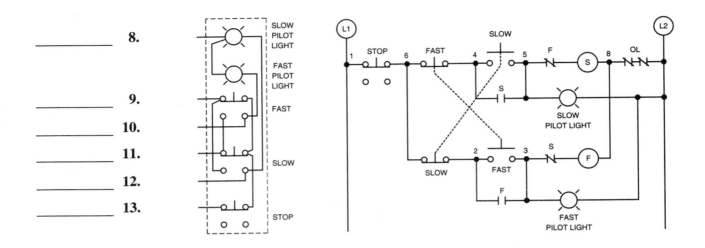

8. _____

9. _____

10. _____

11. _____

12. _____

13. _____

Activity 3-6. Basic Switching Logic

1. Add a second start button to the basic control circuit so Start Button 1 or Start Button 2 can be used to start the motor. Include a second stop button that is connected so that Stop Button 1 or Stop Button 2 can be used to stop the motor.

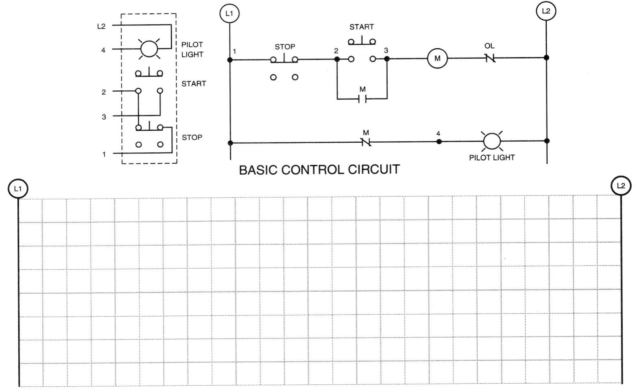

BASIC CONTROL CIRCUIT

BASIC CONTROL CIRCUIT WITH SECOND START BUTTON

2. Add a pressure switch to the basic control circuit to automatically stop the motor when the pressure in the system exceeds a setpoint pressure. Include a temperature switch to automatically stop the motor when the temperature in the system exceeds a set temperature.

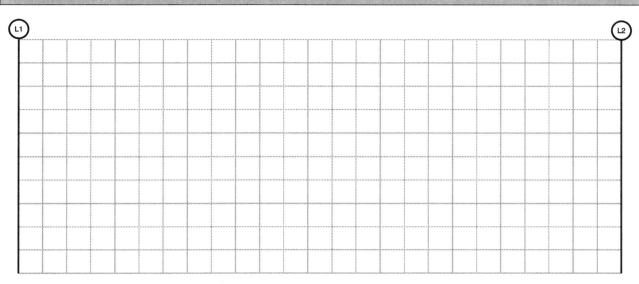

BASIC CONTROL CIRCUIT WITH PRESSURE SWITCH

3. Add a fuse to the basic control circuit with pressure switch to automatically stop the motor when the fuse blows or the overload trips. Include an emergency start button to manually start the motor even when all the other buttons and switches are open (except the overload contacts or fuse). This allows for a manual start even when the temperature and/or pressure is above the setpoint.

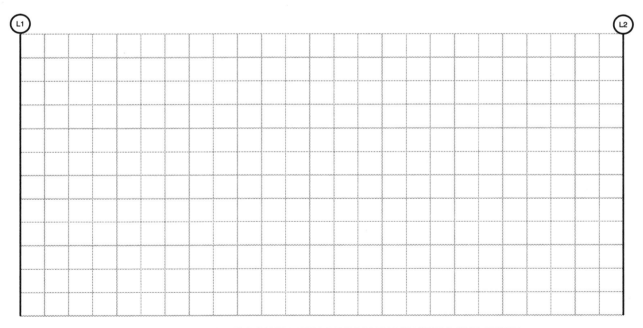

BASIC CONTROL CIRCUIT WITH PRESSURE SWITCH AND FUSE

4. Identify the logic function of each line diagram.

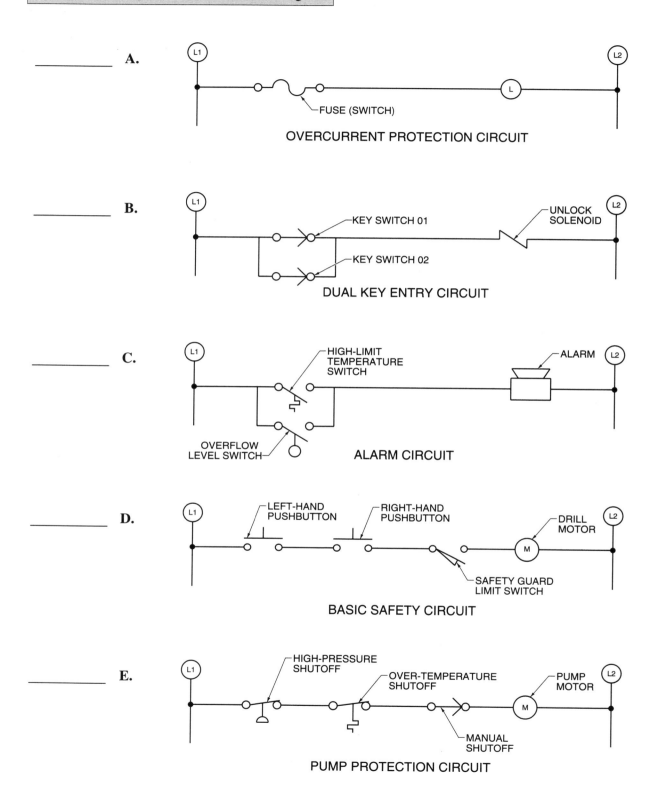

_____ **A.**

FUSE (SWITCH)

OVERCURRENT PROTECTION CIRCUIT

_____ **B.**

KEY SWITCH 01

KEY SWITCH 02

UNLOCK SOLENOID

DUAL KEY ENTRY CIRCUIT

_____ **C.**

HIGH-LIMIT TEMPERATURE SWITCH

ALARM

OVERFLOW LEVEL SWITCH

ALARM CIRCUIT

_____ **D.**

LEFT-HAND PUSHBUTTON

RIGHT-HAND PUSHBUTTON

DRILL MOTOR

SAFETY GUARD LIMIT SWITCH

BASIC SAFETY CIRCUIT

_____ **E.**

HIGH-PRESSURE SHUTOFF

OVER-TEMPERATURE SHUTOFF

PUMP MOTOR

MANUAL SHUTOFF

PUMP PROTECTION CIRCUIT

APPLICATIONS

☐ Application — Motor Coupling Selection

Motor Couplings

A *motor coupling* is a device that connects the motor shaft to the equipment the motor is driving. A motor coupling allows the motor to operate the driven equipment, allows for a slight misalignment between the motor and the driven equipment, and allows for horizontal and axial movement of the shafts.

Misalignment

When connecting equipment, angular misalignment and parallel misalignment occur. *Angular misalignment* is misalignment when two shafts are not parallel. *Parallel misalignment* is misalignment when two shafts are parallel but not on the same line. **See Motor Couplings.**

Motor Coupling Rating

Motor couplings are rated according to the amount of torque they can handle. Couplings are rated in inch-pounds (in-lb) or foot-pounds (ft-lb). The coupling torque rating must be correct for the application to prevent the coupling from bending or breaking. A bent coupling causes misalignment and vibration. A broken coupling prevents the motor from doing work.

Selecting Motor Couplings

The correct coupling for an application is selected by determining the nominal torque rating of the power source, determining the application service factor, calculating the coupling torque rating, selecting a coupling with an equal or greater torque rating, and ensuring that the coupling has the correct shaft size to fit the drive unit.

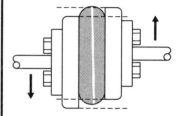

PARALLEL MISALIGNMENT

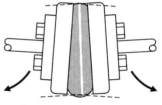

ANGULAR MISALIGNMENT

To select the correct coupling for an application, apply the procedure:

Step 1. Determine the nominal torque rating of the power source (electric motor or other power source). The nominal torque rating is calculated or found on a conversion table.

Finding Nominal Torque Rating — Calculation

To calculate the nominal torque rating of a motor in in-lb, apply the formula:

$$T = \frac{HP \times 63,000}{rpm}$$

where

T = nominal torque rating (in in-lb)　　　　　63,000 = constant
HP = horsepower　　　　　　　　　　　　　　rpm = speed (in revolutions per minute)

To calculate the nominal torque rating of a motor in ft-lb, apply the formula:

$$T = \frac{HP \times 5252}{rpm}$$

where

T = nominal torque rating (in ft-lb)　　　　　5252 = constant
HP = horsepower　　　　　　　　　　　　　　rpm = speed (in revolutions per minute)

Example: Finding Nominal Torque Rating — Calculation

A 5 HP motor operates at 1740 rpm. Find the nominal torque rating of the motor in ft-lb.

$$T = \frac{HP \times 5252}{rpm}$$

$$T = \frac{5 \times 5252}{1740}$$

$$T = \frac{26,260}{1740}$$

$$T = \textbf{15.09 ft-lb}$$

Note: 15.09 ft-lb = 181.08 in-lb (15.09 × 12)

CONVERSION FACTOR		
Multiply	**By**	**To obtain**
ft-lb	12	in-lb
Divide	**By**	**To obtain**
in-lb	12	ft-lb

Finding Nominal Torque Rating — Conversion Table

To find the nominal torque rating of a motor using a conversion table, place one end of a straightedge on the rpm and the other end on the horsepower. The point where the straightedge crosses the torque scale is the torque rating. **See Horsepower To Torque Conversion in Appendix.**

Example: Finding Nominal Torque Rating — Conversion Table

A 1 HP motor operates at 1750 rpm. Find the nominal torque rating in ft-lb.

Place one end of a straightedge on 1750 rpm and the other end at 1 HP. The straightedge crosses the torque scale at 3 ft-lb.

Step 2. Determine application service factor. An *application service factor* is a multiplier that corrects for the operating conditions of the coupling. The greater the stress placed on the coupling, the larger the multiplier. By applying a multiplier, the size of the coupling is increased to adjust for severity of the load placed on the motor. **See Common Service Factors in Appendix.**

Step 3. Calculate the coupling torque rating by multiplying the nominal torque rating of the power source by the service factor of the application. Coupling torque rating is found by applying the formula:

$$C_{TR} = N_{TR} \times SF$$

where

C_{TR} = coupling torque rating (in in-lb or ft-lb)

N_{TR} = nominal torque rating (in in-lb or ft-lb)

SF = service factor

Example: Finding Coupling Torque Rating

The nominal torque rating of a motor is 16 ft-lb. The motor/coupling application is a concrete mixer. Find the coupling torque rating.

$$C_{TR} = N_{TR} \times SF$$

$$C_{TR} = 16 \times 2 \text{ (from Common Service Factors in Appendix)}$$

$$C_{TR} = \textbf{32 ft-lb}$$

Note: 32 ft-lb = 384 in-lb

Step 4. Select a coupling with an equal or greater torque rating. **See Coupling Selections in Appendix.**

Step 5. Ensure coupling has the correct shaft size to fit the drive unit. The exact size of a motor shaft can be determined by the motor frame number. Typical shaft sizes for motors from ¼ HP to 200 HP are ½″, ⅝″, ⅞″, 1⅛″, 1⅜″, 1⅝″, 1⅞″, 2⅛″, 2⅜″, 2⅞″, and 3⅜″.

Example: Motor Coupling Selection

A 1.5 HP motor operating at 1750 rpm is used in a heavy-duty conveyor. Select the coupling for the application.

Step 1. Determine nominal torque rating.

$$T = \frac{HP \times 63{,}000}{rpm}$$

$$T = \frac{1.5 \times 63{,}000}{1750}$$

$$T = \frac{94{,}500}{1750}$$

$T = \textbf{54 in-lb}$

Note: 54 in-lb = 4.5 ft-lb (54 ÷ 12)

Step 2. Determine application service factor.

From Common Service Factors in Appendix, the service factor for a heavy-duty conveyor is 2.

Step 3. Calculate the coupling torque rating.

$C_{TR} = N_{TR} \times SF$
$C_{TR} = 54 \times 2$
$C_{TR} = \textbf{108 in-lb}$

Step 4. Select a coupling with an equal or greater torque rating.

From Coupling Selections in Appendix, the coupling with a torque rating equal to or greater than 108 in-lb is a 10-104-A.

Step 5. Ensure coupling has the correct shaft size to fit the drive unit.

A 1.5 HP motor normally has a 145T frame. The shaft size of a 145T frame motor is ⅞″. The coupling must have a bore size that would accept the ⅞″ motor shaft.

Application — Wiring Diagrams

A *wiring diagram* is a diagram that shows the placement and connections of all components in the control circuit and power circuit. When working with wiring diagrams, it is difficult to see the circuit operation because of the number of wires. To better understand the circuit operation, a line diagram is used. A line diagram can be drawn from the wiring diagram by tracing each wire and drawing them in line diagram form. A wiring diagram is used when troubleshooting, servicing, or repairing an operating circuit. **See Wiring Diagram.**

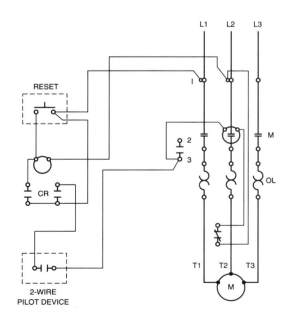

RESET

CR

2-WIRE
PILOT DEVICE

L1 L2 L3

2
3

M

OL

T1 T2 T3

M

Application — Ordering Replacement Parts

A manufacturer's service bulletin is used to order replacement parts. A *service bulletin* is a drawing and parts list of a device. The service bulletin contains the part number of the component requiring replacement. Service bulletins should be kept for each piece of equipment and are included when the equipment is purchased. If a service bulletin is not included, one may be obtained from the manufacturer. **See Service Bulletin in Appendix.**

Application — Enclosure Selection

An *enclosure* is a device that protects a motor starter and control devices. Enclosures are categorized by the protection they provide. An enclosure is selected based on the location of the equipment and the NEC® requirements. **See Motor Enclosures.**

MOTOR ENCLOSURES				
Type	Use	Service conditions	Tests	Comments
1	Indoor	No unusual	Rod entry, rust resistance	
3	Outdoor	Windblown dust, rain, sleet, and ice on enclosure	Rain, external icing, dust, and rust resistance	Do not provide protection against internal condensation or internal icing
3R	Outdoor	Falling rain and ice on enclosure	Rod entry, rain, external icing, and rust resistance	Do not provide protection against dust, internal condensation, or internal icing
4	Indoor/outdoor	Windblown dust and rain, splashing water, hose-directed water, and ice on enclosure	Hosedown, external icing, and rust resistance	Do not provide protection against internal condensation or internal icing
4X	Indoor/outdoor	Corrosion, windblown dust and rain, splashing water, hose-directed water, and ice on enclosure	Hosedown, external icing, and corrosion resistance	Do not provide protection against internal condensation or internal icing
6	Indoor/outdoor	Occasional temporary submersion at a limited depth		
6P	Indoor/outdoor	Prolonged submersion at a limited depth		
7	Indoor locations classified as Class I, Groups A, B, C, or D, as defined in the NEC®	Withstand and contain an internal explosion of specified gases, contain an explosion sufficiently so an explosive gas-air mixture in the atmosphere is not ignited	Explosion, hydrostatic, and temperature	Enclosed heat-generating devices shall not cause external surfaces to reach temperatures capable of igniting explosive gas-air mixtures in the atmosphere

continued

continued

\		**MOTOR ENCLOSURES**		
Type	**Use**	**Service conditions**	**Tests**	**Comments**
9	Indoor locations classified as Class II, Groups E or G, as defined in the NEC®	Dust	Dust penetration, temperature, and gasket aging	Enclosed heat-generating devices shall not cause external surfaces to reach temperatures capable of igniting explosive gas-air mixtures in the atmosphere
12	Indoor	Dust, falling dirt, and dripping noncorrosive liquids	Drip, dust, and rust resistance	Do not provide protection against internal condensation
13	Indoor	Dust, spraying water, oil, and noncorrosive coolant	Oil explosion and rust resistance	Do not provide protection against internal condensation

The NEC® classifies hazardous locations according to the properties and quantities of the hazardous material that may be present. Hazardous locations are divided into three classes, two divisions, and seven groups. *Class* refers to the generic hazardous material present. Class I applies to locations where flammable gases or vapors may be present in the air in quantities sufficient to produce an explosive or ignitable mixture. Class II applies to locations where combustible dusts may be present in sufficient quantity to cause an explosion. Class III applies to locations where the hazardous material consists of easily ignitable fibers or flyings that are not normally in suspension in the air in large enough quantities to produce an ignitable mixture.

Division applies to the probability that a hazardous material is present. Division 1 applies to locations where ignitable mixtures exist under normal operating conditions found in the process, operation, or during periodic maintenance. Division 2 applies to locations where ignitable mixtures exist only in abnormal situations. Abnormal situations occur as a result of accidents or when equipment fails.

Air mixtures of gases, vapors, and dusts are grouped according to their similar characteristics. The NEC® classifies gases and vapors in Groups A, B, C, and D for Class I locations and combustible dusts in Groups E, F, and G for Class II locations. **See Hazardous Locations.**

For example, a type 7 enclosure is required for an indoor application where gasoline is stored (Class I, Group D).

		HAZARDOUS LOCATIONS
Class	**Group**	**Material**
I	A	Acetylene
	B	Hydrogen, Butadiene, ethylene oxide, propylene oxide
	C	Carbon monoxide, ether, ethylene, hydrogen sulfide, morpholine, cyclopropane
	D	Gasoline, benzene, butane, propane, alcohol, acetone, ammonia, vinyl chloride
II	E	Metal dusts
	F	Carbon black, coke dust, coal
	G	Grain dust, flour, starch, sugar, plastics
III	No groups	Wood chips, cotton, flax, and nylon

Application — Sizing Motor Protection, Starter, and Wire — 1φ Motors

When a motor is energized, a high inrush current occurs. This inrush current is many times the normal running current of the motor. Motors require special overload protection devices that can withstand the high starting current and still protect the motor from a sustained overload. Dual-element fuses (Fusetron®) are used to protect motors from overcurrent and short circuits.

Manufacturers provide charts for selecting the correct size and type of overload protection for a motor. These charts include information such as sizing back-up protection, switch size, starter size, temperature rating, and wire and conduit sizes. **See 1φ Motors and Circuits and Appendix.**

1φ Motors

The 1φ motor chart is used with 115 V and 230 V 1φ motors up to 10 HP. Column 1 lists the horsepower and ampere rating for motors operating at normal speeds. Column 2 lists the fuse size for different motor temperature ratings and service factors. Column 3 lists the switch or fuse holder size. The size listed denotes the minimum ampere rating. Column 4 lists the minimum motor starter required when controlling the motor. The motor starter usually provides overload protection in addition to controlling the motor. Columns 5 and 6 list the controller termination temperature rating and minimum size copper wire and conduit required when connecting the motor. A bullet in Column 5 denotes insulation available. Column 6 is used in conjunction with Column 5 because wire type (TW, THW, etc.) and wire size must be considered.

1φ MOTORS AND CIRCUITS											
1	**2**		**3**	**4**	**5**				**6**	
Size of motor Table 430-148	**Motor overload protection** Low-peak or Fusetron®				**Controller termination temperature rating**				**Minimum size of copper wire and trade conduit**	
					60°C		**75°C**			
	Motor less than 40°C or greater than 1.15 SF (Max fuse 125%)	**All other motors (Max fuse 115%)**	**Switch 115% minimum or HP rated or fuse holder size**	**Minimum size of starter**					**Wire size (AWG or kcmil)**	**Conduit (in.)**
HP Amp					**TW**	**THW**	**TW**	**THW**		
115 V (120 V system)										
⅙ 4.4	5	5	30	00	•	•	•	•	14	½
¼ 5.8	7	6¼	30	00	•	•	•	•	14	½
⅓ 7.2	9	8	30	0	•	•	•	•	14	½
½ 9.8	12	10	30	0	•	•	•	•	14	½
¾ 13.8	15	15	30	0	•	•	•	•	14	½
1 16	20	17½	30	0	•	•	•	•	14	½

3φ Motors

The 3φ motor chart is used with 230 V and 460 V 3φ motors up to 300 HP. Once the motor size is determined, the chart is used in the same manner as the 1φ motor chart. **See 3φ Motors and Circuits in Appendix.**

Direct Current (DC) Motors

The direct current motor chart is used with 90 V, 120 V, and 180 VDC motors up to 10 HP. Once the motor size is determined, the chart is used in the same manner as the 1φ motor chart. **See DC Motors and Circuits in Appendix.**

ACTIVITIES

Name _____ Date _____

⬤ Activity 4-1. Motor Coupling Selection

Calculate the coupling torque rating and select a coupling using Coupling Selections in Appendix.

_____ **1.** A heavy-duty conveyor has a 2 HP motor that turns at 1800 rpm. The coupling torque rating is _____ in-lb.

_____ **2.** A(n) _____ coupling is used for an application with a 205 in-lb torque rating.

_____ **3.** A uniformly loaded standard conveyor has a 5 HP motor that turns at 1200 rpm. The coupling torque rating is _____ in-lb.

_____ **4.** A printing press has an 8.25 HP motor that turns at 1800 rpm. The coupling torque rating is _____ in-lb.

_____ **5.** A textile loom has a 2.5 HP motor that turns at 900 rpm. The coupling torque rating is _____ ft-lb.

_____ **6.** A punch press has a 3.75 HP motor that turns at 1200 rpm. The coupling torque rating is _____ ft-lb.

_____ **7.** A(n) _____ coupling is used for an application with a 30 in-lb torque rating.

_____ **8.** A(n) _____ coupling is used for an application with a 475 in-lb torque rating.

_____ **9.** A(n) _____ coupling is used for an application with a 154 in-lb torque rating.

_____ **10.** A(n) _____ coupling is used for an application with a 17 in-lb torque rating.

⬤ Activity 4-2. Circuit Wiring — 1φ Motor

1. Draw the line diagram that shows the logic of the control circuit. Use standard lettering, numbering, and coding information.

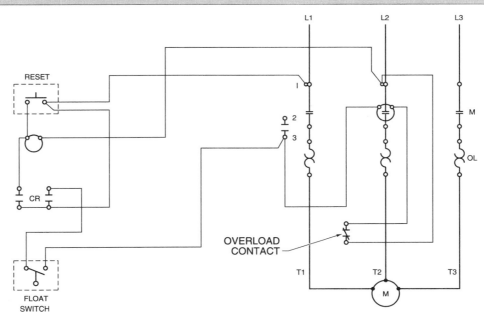

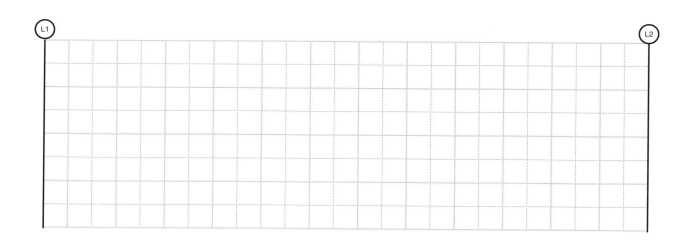

2. Redraw the line diagram adding a pressure switch that automatically stops the motor if pressure increases too high.

3. Redraw the line diagram adding a pilot light that indicates the motor is energized.

Activity 4-3. Circuit Wiring — Remote Pushbutton Station

1. Draw the line diagram that shows the logic of the control circuit. Use standard lettering, numbering, and coding information.

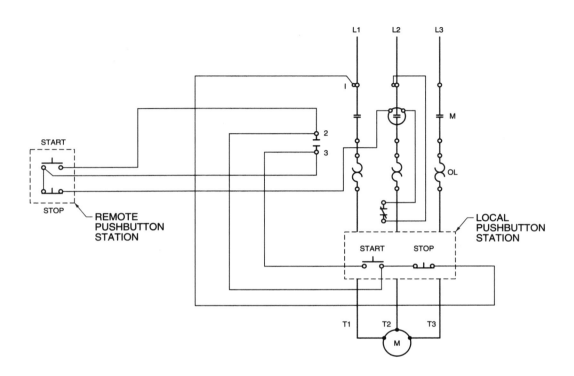

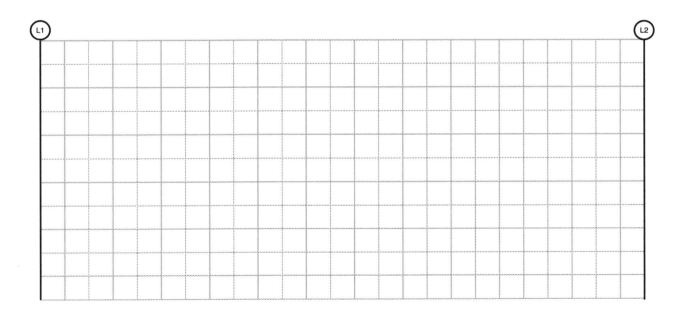

2. Redraw the line diagram adding a third remote start pushbutton that starts the motor.

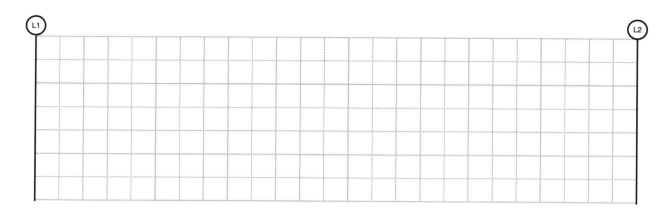

3. Redraw the line diagram adding a third remote stop pushbutton that stops the motor.

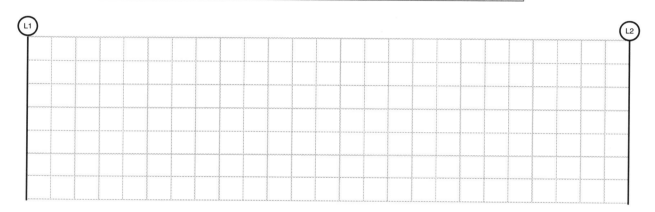

4. Redraw the line diagram adding a two-position (jog/run) selector switch. When the selector switch is in the jog position, the three start pushbuttons jog the motor.

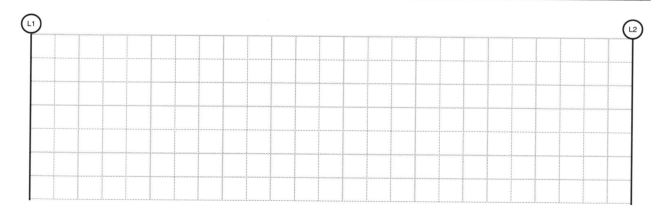

◖ Activity 4-4.   Circuit Wiring — Sequential Motor Starting

1. Draw the line diagram that shows the logic of the control circuit. Use standard lettering, numbering, and coding information.

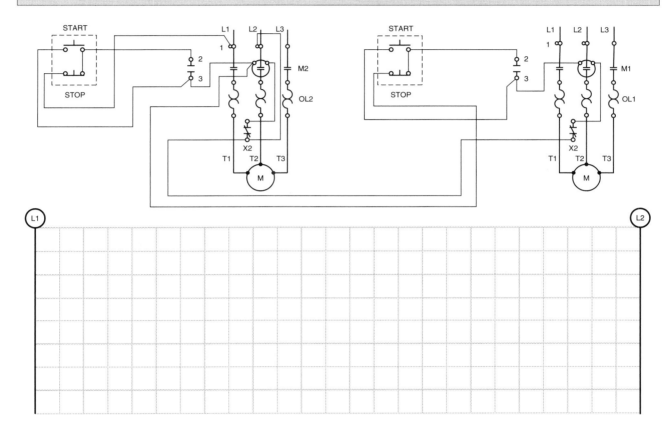

2. Redraw the line diagram adding a manual override pushbutton that starts Motor 1 when Motor 2 is not running. The manual override pushbutton runs the motor when pressed and held as long as Motor 2 is not running.

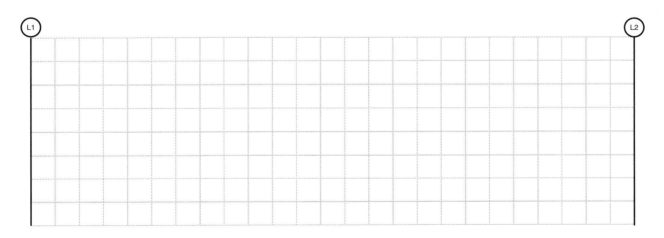

3. Redraw the line diagram adding a limit switch that turns OFF both motors if the switch is pressed.

L1   L2

◯ Activity 4-5. Ordering Replacement Parts

Answer the questions using Service Bulletin in Appendix.

_____ **1.** The part number used to replace a normally closed internal interlock is _____.

_____ **2.** The part number of the movable contact is _____.

_____ **3.** The part number(s) of the clamp that holds the incoming power lines and motor terminal wires is _____.

_____ **4.** _____ clamps are required for the starter when controlling a 3φ motor.

_____ **5.** What is the difference between items B and C?

◯ Activity 4-6. Selecting a Motor Enclosure

List the enclosure that best fits the application using Motor Enclosures on page 32.

_____ **1.** A type _____ enclosure is used for a commercial basement that occasionally floods.

_____ **2.** A type _____ enclosure is used for a warehouse used to store paper.

_____ **3.** A type _____ enclosure is used in an industrial Class II, Group E location.

_____ **4.** A type _____ enclosure is used in an industrial Class I, Group B location.

_____ **5.** A type _____ enclosure is used in an industrial building in which the enclosure is occasionally hosed down to clean a noncorrosive coolant.

◯ Activity 4-7. Sizing Motor Protection, Starter, and Wire — 1φ Motors

List the devices based on the motor installation data. Use 1φ Motors and Circuits in Appendix.

1. A 1 HP, 230 V motor with a 40°C rating and a 1.15 SF is installed in a 60°C (or less) location.

_____ **A.** The motor overload protection fuse is _____ A.

_____ **B.** The motor starter is size _____.

_____   **C.** The wire size is No. _____.

_____   **D.** The conduit size is _____″.

2. A ½ HP, 230 V motor with a 40°C rating and a 1.15 SF is installed in a 60°C (or less) location.

_____   **A.** The motor overload protection fuse is _____ A.

_____   **B.** The motor starter is size _____.

_____   **C.** The wire size is No. _____.

_____   **D.** The conduit size is _____″.

3. A ¾ HP, 115 V motor with a 40°C rating and a 1.15 SF is installed in a 60°C (or less) location.

_____   **A.** The motor overload protection fuse is _____ A.

_____   **B.** The motor starter is size _____.

_____   **C.** The wire size is No. _____.

_____   **D.** The conduit size is _____″.

4. A 2 HP, 115 V motor with a 50°C rating and a 1.15 SF is installed in a 60°C (or less) location.

_____   **A.** The motor overload protection fuse is _____ A.

_____   **B.** The motor starter is size _____.

_____   **C.** The wire size is No. _____.

_____   **D.** The conduit size is _____″.

⬤ Activity 4-8. Sizing Motor Protection, Starter, and Wire — 3φ Motors

List the devices based on the motor installation data. Use 3φ Motors and Circuits in Appendix.

1. A 3 HP, 230 V motor with a 40°C rating and a 1.15 SF is installed in a 60°C (or less) location.

_____   **A.** The motor overload protection fuse is _____ A.

_____   **B.** The motor starter is size _____.

_____   **C.** The wire size is No. _____.

_____   **D.** The conduit size is _____″.

2. A 75 HP, 230 V motor with a 40°C rating and a 1.15 SF is installed in a 60°C (or less) location.

_____   **A.** The motor overload protection fuse is _____ A.

_____   **B.** The motor starter is size _____.

_____   **C.** The wire size is No. _____.

_____   **D.** The conduit size is _____″.

3. A 40 HP, 460 V motor with a 40°C rating and a 1.15 SF is installed in a 60°C (or less) location.

_____   **A.** The motor overload protection fuse is _____ A.

_____   **B.** The motor starter is size _____.

_____   **C.** The wire size is No. _____.

_____   **D.** The conduit size is _____″.

4. A 150 HP, 460 V motor with a 50°C rating and a 1.15 SF is installed in a 60°C (or less) location.

_____   **A.** The motor overload protection fuse is _____ A.

_____   **B.** The motor starter is size _____.
_____   **C.** The wire size is No. _____.
_____   **D.** The conduit size is _____″.

◯ Activity 4-9.   Sizing Motor Protection, Starter, and Wire — DC Motors

List the devices based on the motor installation data. Use DC Motors and Circuits in Appendix.

1. A 1 HP, 120 V motor with a 40°C rating and a 1.15 SF is installed in a 60°C (or less) location.
_____   **A.** The motor overload protection fuse is _____ A.
_____   **B.** The motor starter is size _____.
_____   **C.** The wire size is No. _____.
_____   **D.** The conduit size is _____″.

2. A 10 HP, 120 V motor with a 40°C rating and a 1.15 SF is installed in a 60°C (or less) location.
_____   **A.** The motor overload protection fuse is _____ A.
_____   **B.** The motor starter is size _____.
_____   **C.** The wire size is No. _____.
_____   **D.** The conduit size is _____″.

3. A 3 HP, 180 V motor with a 40°C rating and a 1.15 SF is installed in a 60°C (or less) location.
_____   **A.** The motor overload protection fuse is _____ A.
_____   **B.** The motor starter is size _____.
_____   **C.** The wire size is No. _____.
_____   **D.** The conduit size is _____″.

4. A 5 HP, 180 V motor with a 50°C rating and a 1.15 SF is installed in a 60°C (or less) location.
_____   **A.** The motor overload protection fuse is _____ A.
_____   **B.** The motor starter is size _____.
_____   **C.** The wire size is No. _____.
_____   **D.** The conduit size is _____″.

APPLICATIONS

Application — Fluid Power Color Coding

The three basic types of power used to produce work are mechanical (shafts, gears, belts), electrical (motors, solenoids), and fluid power (hydraulic or pneumatic). **See Power Methods.** Symbols are used for designing, operating, and maintaining fluid power systems. Fluid power symbols make the different components in the circuit easier to understand. **See Selected Symbols. See Fluid Power Symbols in Appendix.** Fluid power is used in industrial applications because it develops a large amount of power in a small space. Fluid power circuits use color codes to identify operating conditions in a circuit. **See Fluid Power Color Codes.**

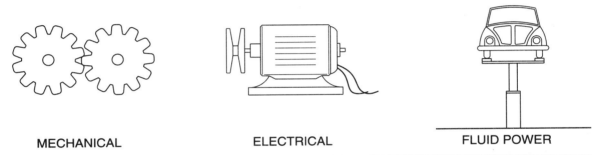

MECHANICAL   ELECTRICAL   FLUID POWER

SELECTED SYMBOLS				
CYLINDERS		**HYDRAULIC PUMPS**		**HYDRAULIC MOTORS**
DOUBLE-ACTING	DOUBLE-ACTING W/ DOUBLE END ROD	FIXED DISPLACEMENT — 1 ROTATION	VARIABLE DISPLACEMENT — 1 ROTATION	FIXED DISPLACEMENT — 1 ROTATION / DUAL ROTATION
DIRECTIONAL CONTROL VALVES		**FLOW CONTROL VALVE**	**PRESSURE CONTROL VALVE**	
TWO-WAY NC	FOUR-WAY SOLENOID OPERATED			
AIR OR OIL FILTER	**HEAT EXCHANGER**	**PRESSURE GAUGE**	**CHECK VALVE**	**MANUAL SHUT-OFF**

FLUID POWER COLOR CODES			
Color	Meaning	Color	Meaning
Red	System operating pressure or highest working pressure. *System pressure* is the fluid flow after the pump until the flow is reduced, metered, or returned to the tank.	Yellow	Controlled flow by a metering device. *Controlled flow* is the fluid flow after a flow control valve has reduced the volume (gpm) of fluid flow.
Blue	Exhaust or return flow back to the tank. *Exhaust flow* is the fluid flow from the actuator, back through the valve, and to the tank.	Orange	Reduced pressure that is lower than system operating pressure. *Reduced flow* is the fluid flow after a pressure-reducing valve has reduced the pressure (psi) of the fluid.
Green	Intake flow to pump or drain line flow. *Intake flow* is the fluid flow from the reservoir tank, through the filters, and to the pump.	Violet	Intensified pressure. *Intensified pressure* is pressure higher than system operating pressure.
		None	Inactive hydraulic fluid (reservoir fluid).

▢ Application — Valve Selection

Directional Control Valves

Directional control valves control the direction of fluid flow in a fluid power system. Directional control valves do not affect the pressure or flow rate of the fluid. Directional control valves are identified by the number of positions, ways, and type of actuators.

Positions

Directional control valves are placed in different positions to start, stop, or change the direction of fluid flow. A *position* is the number of positions within the valve that the spool is placed in to direct fluid flow through the valve. Two- and three-way valves have two positions. Four-way valves may have two or three positions. **See Valves.**

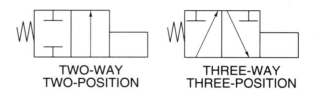

TWO-WAY
TWO-POSITION

THREE-WAY
THREE-POSITION

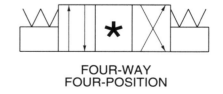

FOUR-WAY
FOUR-POSITION

Ways

A *way* is the flow path through the valve. Most directional control valves are either two-way, three-way, or four-way. The number of ways required depends on the application. Two-way directional control valves have two main ports that allow or stop the flow of fluid. Two-way valves are used as shutoff, check, and quick exhaust valves. Three-way valves allow or stop fluid flow or exhaust. They are used to control single-acting cylinders, fill and drain tanks, and control nonreversible fluid motors. Four-way directional control valves have four (or five) main ports that change fluid flow from one port to another. Four-way valves are used to control the direction of double-acting cylinders or reversible fluid motors.

Actuators

Directional control valves must have means to change the valve position. An *actuator* is a device that changes the valve position. Directional control valve actuators include pilots, solenoids, springs, manual levers, and palm buttons. **See Actuators.**

Solenoid-actuated Valves

Solenoid-actuated valves are used in most fluid power circuits to control the direction of fluid flow, the rate of fluid flow, and the sequence of operation. When ordering and installing electrically operated fluid power valves, the requirements of the fluid power part and the electrical part of the valve must be known.

Valve Parts

The fluid power part includes the valve type (two-, three-, or four-way), the spool (two or three positions), and the actuator (pilot, solenoid, spring, manual). The electrical part of the valve includes the number of solenoids (single or double) and the coil voltage and type (AC or DC).

Valve Numbering System

Valve manufacturers provide a numbering system to simplify valve selection and ordering. The numbering system consists of letters and numbers that represent the different valve models. **See Valve Selections.**

For example, a model 22AS2-120-1 valve is a four-way, two-position stacking valve with $\frac{1}{8}''$ NPTF ports. The valve is solenoid actuated, solenoid returned, and has a 120 VAC coil voltage.

PILOT AND SOLENOID

RETURN SPRING

MANUAL LEVER

PALM BUTTON

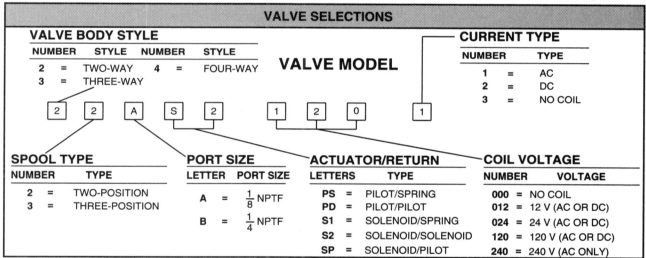

VALVE SELECTIONS

VALVE BODY STYLE

NUMBER	STYLE	NUMBER	STYLE
2 =	TWO-WAY	4 =	FOUR-WAY
3 =	THREE-WAY		

VALVE MODEL

2 2 A S 2 1 2 0 1

CURRENT TYPE

NUMBER	TYPE
1 =	AC
2 =	DC
3 =	NO COIL

SPOOL TYPE

NUMBER	TYPE
2 =	TWO-POSITION
3 =	THREE-POSITION

PORT SIZE

LETTER	PORT SIZE
A =	$\frac{1}{8}$ NPTF
B =	$\frac{1}{4}$ NPTF

ACTUATOR/RETURN

LETTERS	TYPE
PS =	PILOT/SPRING
PD =	PILOT/PILOT
S1 =	SOLENOID/SPRING
S2 =	SOLENOID/SOLENOID
SP =	SOLENOID/PILOT

COIL VOLTAGE

NUMBER	VOLTAGE
000 =	NO COIL
012 =	12 V (AC OR DC)
024 =	24 V (AC OR DC)
120 =	120 V (AC OR DC)
240 =	240 V (AC ONLY)

■ Application — Determining Coil Ratings

Electrical coils use current to produce the power required to move the plunger in a solenoid. The movement of the plunger produces a mechanical force. The mechanical force develops linear motion that is used in many industrial applications.

Coil Specifications

The current drawn depends on the applied voltage and size of the coil. Manufacturers list coil specifications to assist in installation and component sizing. **See Coil Specifications.**

For example, the coil used in a size 2 starter may have two, three, or four poles and draw .25 A sealed current when connected to a 120 V power source. This ampere value is used when selecting the fuse, wire, and transformer sizes.

Size	Number of poles	Inrush current (A) 60 cycles					Sealed current (A) 60 cycles					Approximate operating time (in ms)	
		120 V	208 V	240 V	480 V	600 V	120 V	208 V	240 V	480 V	600 V	Pick-up	Drop-out
00	1-2-3	0.50	0.29	0.25	0.12	0.07	0.12	0.07	0.06	0.03	0.02	28	13
0	1-2-3-4	0.88	0.50	0.44	0.22	0.17	0.14	0.08	0.07	0.04	0.03	29	14
1	1-2-3-4	1.54	0.89	0.77	0.39	0.31	0.18	0.10	0.09	0.04	0.04	26	17
2	2-3-4	1.80	1.04	0.90	0.45	0.36	0.25	0.14	0.13	0.06	0.05	32	14
3	2-3	4.82	2.78	2.41	1.21	0.97	0.36	0.21	0.18	0.09	0.07	35	18
	4	5.34	3.08	2.67	1.33	1.07	0.39	0.23	0.20	0.10	0.08	35	18
4	2-3	8.30	4.80	4.15	2.08	1.66	0.54	0.31	0.27	0.14	0.11	41	18
	4	9.90	5.71	4.95	2.47	1.98	0.61	0.35	0.31	0.15	0.12	41	18
5	2-3	16.23	9.36	8.11	4.06	3.25	0.81	0.47	0.41	0.20	0.16	43	18

COIL SPECIFICATIONS

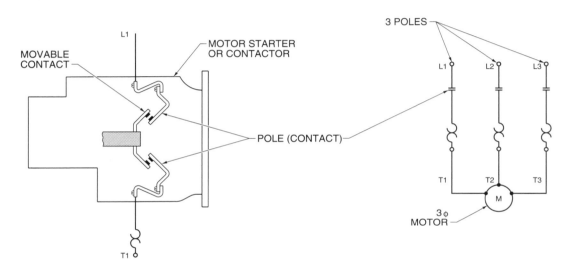

Application — Solenoid in Lubrication Systems

Solenoid-operated valves are used in lubrication systems to start and stop the flow of lubricant. The valve opens when lubricant is required. Closing the valve stops the flow of lubricant. The length of time a valve is open and time between openings depends on the application. **See Chain Lubrication Application.**

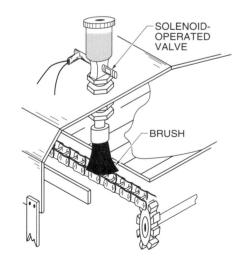

ACTIVITIES

Name _____ Date _____

⬤ Activity 5-1.   Fluid Power Color Coding

Mark the color in each part of the fluid power circuit using Fluid Power Color Codes on page 44.

_____ **1.** Color is _____.
_____ **2.** Color is _____.
_____ **3.** Color is _____.
_____ **4.** Color is _____.
_____ **5.** Color is _____.

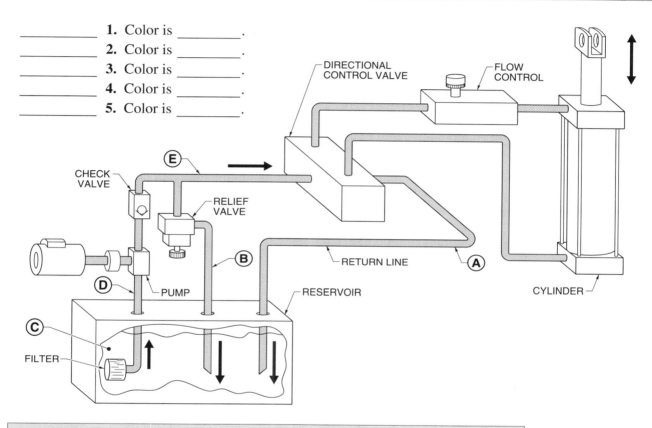

Identify the fluid power symbols on Clamp and Eject Circuit on page 48.

_____ **6.** Two-way, cam-operated valve
_____ **7.** Pressure gauge
_____ **8.** Pressure-reducing valve
_____ **9.** Double-acting cylinder
_____ **10.** Needle valve
_____ **11.** Check valve
_____ **12.** Hydraulic pump
_____ **13.** Pressure-relief valve
_____ **14.** Filter
_____ **15.** Four-way, solenoid-operated valve

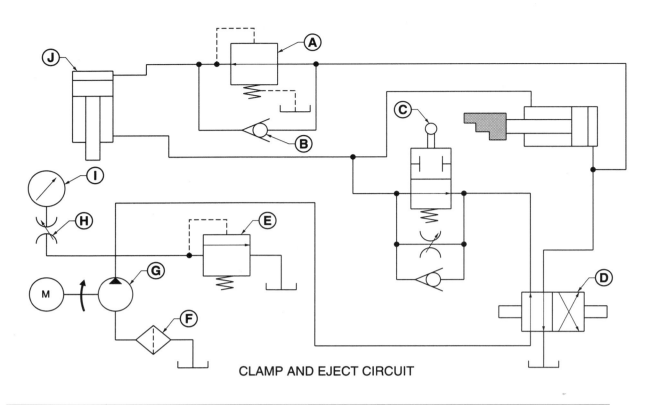

CLAMP AND EJECT CIRCUIT

16. Mark the color in each part of the fluid power circuit using Fluid Power Color Codes on page 44.

_____ **A.** Color is _____.
_____ **B.** Color is _____.
_____ **C.** Color is _____.
_____ **D.** Color is _____.
_____ **E.** Color is _____.

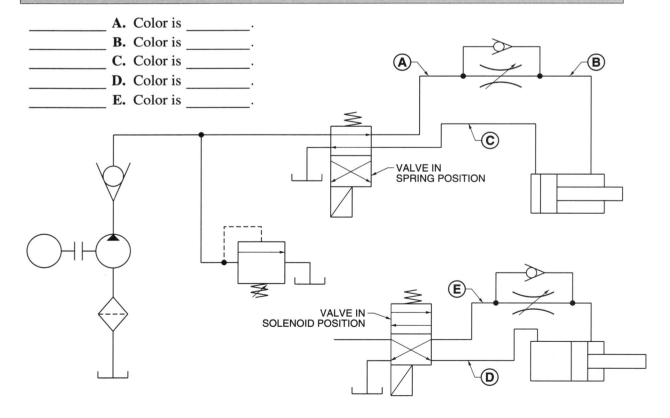

◯ Activity 5-2. Solenoid Valve Ordering

List the model number using Valve Selections on page 45.

_____ **1.** The model number for a three-way, two-position, $1/8''$ NPTF, solenoid-actuated, spring-return, 120 VAC operated valve is _____.

_____ **2.** The model number for a four-way, three-position, $1/8''$ NPTF, solenoid-actuated, solenoid-return, 120 VDC operated valve is _____.

_____ **3.** The model number for a two-way, two-position, $1/4''$ NPTF, solenoid-actuated, pilot-return, 240 VAC operated valve is _____.

_____ **4.** The model number for a three-way, two-position, $1/4''$ NPTF, solenoid-actuated, spring-return, 24 VDC operated valve is _____.

_____ **5.** The model number for a three-way, three-position, $1/4''$ NPTF, solenoid-actuated, solenoid-return, 24 VAC operated valve is _____.

_____ **6.** A model number _____ valve controls a double-acting cylinder. When one pushbutton is pressed and released, the cylinder advances and when a second pushbutton is pressed and released, the cylinder retracts. The circuit is a 120 VAC and requires a $1/8''$ NPTF port.

◯ Activity 5-3. Determining Coil Ratings

Answer the questions using Coil Specifications on page 46.

1. A 120 V, size 00, 3-pole contactor is used for an application.

_____ **A.** The inrush current is _____ A.

_____ **B.** The sealed current is _____ A.

_____ **C.** The pick-up time is _____ ms.

_____ **D.** The drop-out time is _____ ms.

2. A 240 V, size 5, 3-pole contactor is used for an application.

_____ **A.** The inrush current is _____ A.

_____ **B.** The sealed current is _____ mA.

_____ **C.** The pick-up time is _____ ms.

_____ **D.** The drop-out time is _____ ms.

3. A 120 V, size 3, 3-pole contactor is used in an application.

_____ **A.** The inrush current is _____ A.

_____ **B.** The sealed current is _____ mA.

_____ **C.** The pick-up time is _____ ms.

_____ **D.** The drop-out time is _____ ms.

4. A 480 V, size 3, 3-pole contactor is used in an application.

_____ **A.** The inrush current is _____ A.

_____ **B.** The sealed current is _____ A.

_____ **C.** The pick-up time is _____ ms.

_____ **D.** The drop-out time is _____ ms.

◐ Activity 5-4. Solenoid in a Lubrication System

1. Draw the line diagram so lubricant is dispensed every time the pushbutton is pressed and held, and a two-position selector switch is in the hand position.

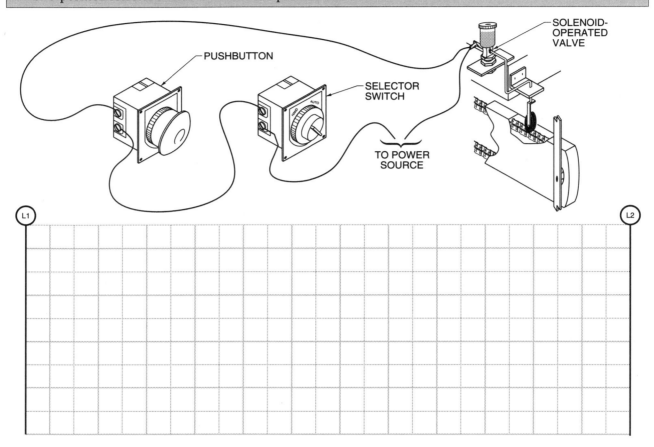

2. Redraw the line diagram so lubricant is dispensed every time the temperature is above a set temperature, and the two-position selector switch is in the automatic position.

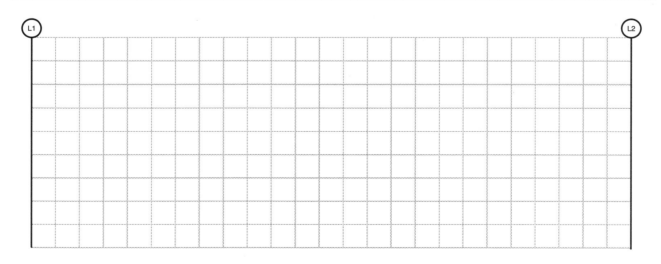

▢ Application — Conveyor Belt Tracking

Conveyor Terminology

The two drive methods used to move product along a conveyor belt are the direct drive and roller drive methods. The *direct drive method* has the product riding directly on the driven belt and is used for light loads. The *roller drive method* has the belt driving rollers on which the product rides. The roller drive method is used for heavy loads. **See Conveyor Drive Methods.**

The right side of a conveyor is the right side when facing the forward direction of material travel. The two ends of a conveyor are identified by their relationship to the forward direction of material travel. The *tail end* of a conveyor is the end where material is fed. The *head end* of a conveyor is the end from which material is discharged. A drive unit is connected to the tail end pulley or head end pulley.

Belt Tension

To prevent slippage, the conveyor belt must have the correct tension. Proper belt tension is accomplished by adjusting the take-up pulley. A *take-up pulley* is a pulley that is used for correcting belt tension and is not connected to a drive.

Belt Tracking

Proper belt tracking depends on the alignment of the pulleys and rollers. A *pulley* is a revolving cylinder connected to a drive. A *roller* is a revolving cylinder not connected to a drive. As a conveyor belt moves, the belt should remain in the center of the pulleys and rollers. When a belt drifts to one side, the belt edge wears or folds up on the conveyor guard.

A conveyor belt drifts toward the side where pulley and roller centers are not parallel and are the closest to each other. Conveyor belt tracking is adjusted by moving the pulley or snub roller. A *snub roller* is a roller not connected to a drive that is used to correct belt alignment. When aligning a belt, all adjustments should be slight with time allowed for the belt to react to the adjustment.

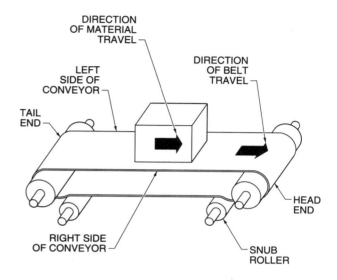

DIRECT DRIVE

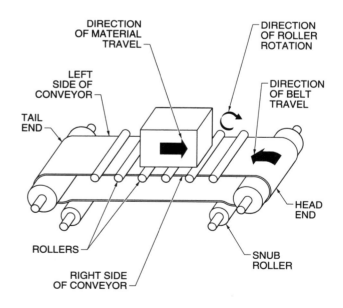

ROLLER DRIVE

Conveyor Adjustment—Head End Pulley

If the belt drifts to the right on the head end pulley during forward material travel, adjust the right side of the head end snub roller in the forward direction of material travel and/or the left side of the head end snub roller in the reverse direction of material travel. **See Conveyors.**

Conveyor Adjustment—Tail End Pulley

If the belt drifts to the right on the tail end pulley during forward material travel, adjust the right side of the tail end snub roller in the reverse direction of material travel and/or the left side of the tail end snub roller in the forward direction of material travel.

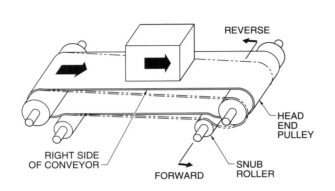

HEAD END PULLEY

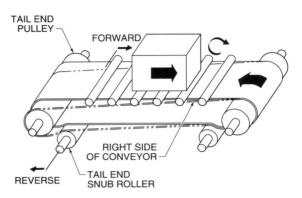

TAIL END PULLEY

Conveyor Adjustment—Center Drive

If the belt drifts to the right side of the center drive and take-up pulleys, adjust the right side of the snub roller in the reverse direction of material travel and/or the left side of the snub roller in the forward direction of material travel. **See Center Drive Conveyors.**

Conveyor Adjustment—Reverse-running Center Drive

If the belt drifts to the right side of the center drive and take-up pulleys, adjust the right side of the snub roller in the forward direction of material travel and/or the left side of the snub roller in the reverse direction.

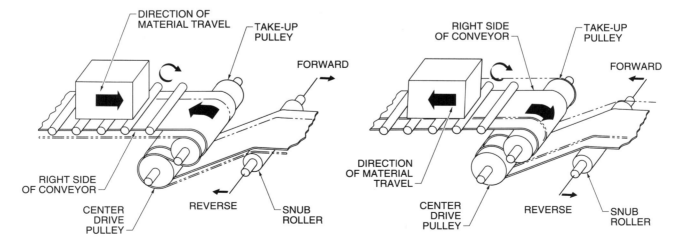

CENTER DRIVE

REVERSE-RUNNING CENTER DRIVE

Application — Dual-voltage Control Transformer Installation

In most industrial applications, the motor power circuit is at a high voltage that produces the required power at the motor. The control circuit is at a low voltage because it is safer. A control transformer reduces the high voltage of the power circuit to a low voltage for the control circuit.

Most control transformers have a dual-voltage primary. A *dual-voltage primary* allows the transformer to be connected to a 240 V or 480 V power circuit. The primary coils of the transformer are connected in series for high voltage and in parallel for low voltage. The secondary side of the transformer delivers a low output (normally 120 V) for the control circuit. **See Dual-voltage Transformer.**

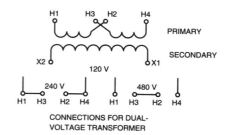

CONNECTIONS FOR DUAL-VOLTAGE TRANSFORMER

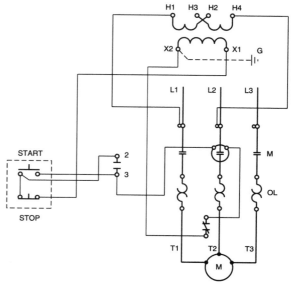

DUAL-VOLTAGE TRANSFORMER

Application — Ambient Temperature Compensation with Overloads

Thermal overloads are heat sensing devices that provide a means of monitoring the current drawn by a motor. When the heat generated by motor windings approaches a damaging level, the overloads trip. In applications where the ambient temperature varies above or below the standard rating temperature of 104°F, ambient temperature must be accounted for because thermal overloads are temperature dependent. If the ambient temperature of the motor is different than that of the overloads, the overloads can cause nuisance tripping or motor burnout.

For example, if nontemperature-compensated overloads are used for a well pump motor that is at a different temperature than the surface-mounted overloads, the overloads must be adjusted for the difference in temperature. Graphs show temperature adjustments for different ambient temperatures. **See Heater Ambient Temperature Correction.**

To find the overload trip current using an ambient temperature correction, apply the procedure:

Step 1. Determine ambient temperature.

Step 2. Find the rated current (%) for the ambient temperature on the graph.

Step 3. Multiply the motor full load current (from motor nameplate) by the rated current.

Example: Finding Overload Trip Current

A motor has a full load current of 10 A. The ambient temperature is 140°F. Find the overload trip current.

Step 1. Determine ambient temperature.

Ambient temperature is 140°F.

Step 2. Find the rated current (%) for the ambient temperature on the graph.

At 140°F the rated current is .87 (from graph).

Step 3. Multiply the motor full load current (from motor nameplate) by the rated current.

Overload trip current = 10 × .87

Overload trip current = **8.7 A**

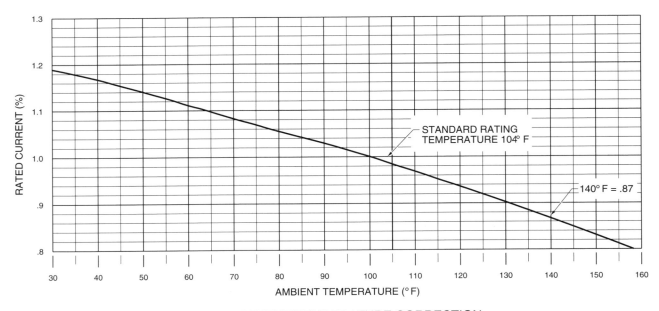

HEATER AMBIENT TEMPERATURE CORRECTION

▮ Application — Determining Overload Trip Time

Overload heaters are designed to trip and disconnect the motor from the power lines when an overload condition occurs. The overload trip time depends on the extent of the overload. The higher the overload, the faster the trip time. Manufacturers provide heater overload trip characteristics. **See Heater Trip Characteristics in Appendix.**

For example, if a motor is overloaded 300%, the overloads trip in 55 seconds (from Heater Trip Characteristics).

▮ Application — Determining Overload Heater Size

The current rating on a motor nameplate is used to select overload heaters. If the exact current rating is not known, the motor full load current is used to find the current rating. **See Full Load Currents in Appendix.**

Manufacturers provide overload heater selections with the motor starter. The heater number is found by matching the motor nameplate-rated current value to the size of the starter. **See Heater Selections in Appendix.**

For example, if a size 1 starter is used to control a motor with a 9 A rating, a #51 heater is required (from Heater Selections).

▮ Application — Contactors

Contactors and motor starters (contactor with added overloads) are rated according to the power and current they switch. Manufacturers provide ratings of different size starters. **See Control Ratings in Appendix.**

ACTIVITIES

Name _____ Date _____

⬤ Activity 6-1. Belt Tracking

Identify the roller that must be adjusted for proper belt tracking. Show the direction of adjustment on the drawing.

_____ **1.** Roller to be adjusted. _____ **2.** Roller to be adjusted.

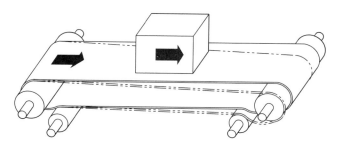
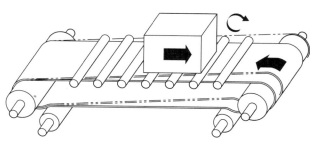

_____ **3.** Roller to be adjusted. _____ **4.** Roller to be adjusted.

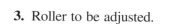
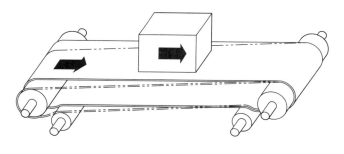
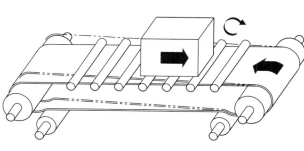

_____ **5.** Roller to be adjusted. _____ **6.** Roller to be adjusted.

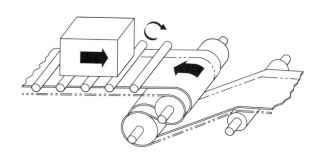
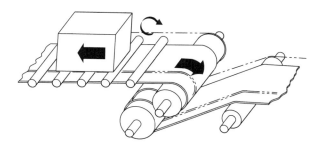

Activity 6-2. Dual-voltage Control Transformer Installation

1. Draw the line diagram that shows the logic of the control circuit. Use standard lettering, numbering, and coding information.

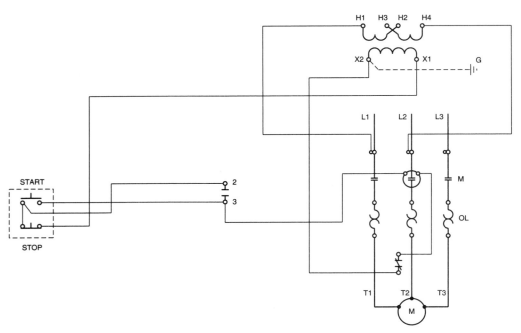

2. Connect the transformer for 240 V to 120 V.

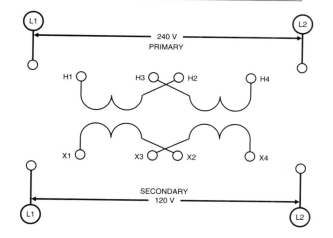

3. Connect the transformer for 240 V to 240 V.

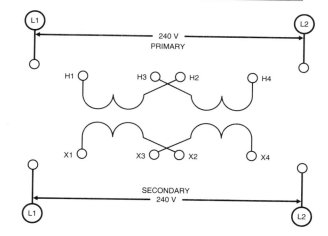

4. Connect the transformer for 480 V to 120 V.

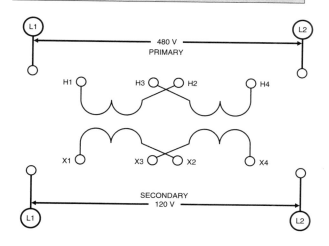

5. Connect the transformer for 480 V to 240 V.

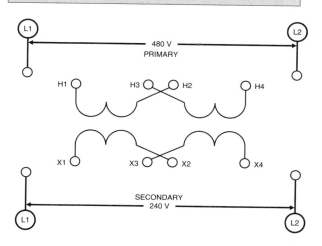

Activity 6-3.   Ambient Temperature Compensation with Overloads

State the overload trip current for each motor installation using the Heater Ambient Temperature Correction on page 54. *Note:* Each motor is installed in a standard enclosure.

_____   **1.** A 25 A rated motor is installed in a 104°F ambient temperature. The overload trip current is _____ A.

_____   **2.** A 35 A rated motor is installed in a 50°F ambient temperature. The overload trip current is _____ A.

_____   **3.** A 50 A rated motor is installed in a 140°F ambient temperature. The overload trip current is _____ A.

_____   **4.** A 75 A rated motor is installed in a 77°F ambient temperature. The overload trip current is _____ A.

_____   **5.** A 12.5 A rated motor is installed in a 149°F ambient temperature. The overload trip current is _____ A.

◯ Activity 6-4. Determining Overload Trip Time

Answer the questions using Heater Trip Characteristics in Appendix.

_____ **1.** If a motor draws 100% of rated current, the overloads _____.
_____ **2.** If a motor draws 500% of rated current, the overloads will trip in _____ seconds.
_____ **3.** If a motor draws 50% of rated current, the overloads _____.
_____ **4.** If a motor draws 1000% of rated current, the overloads will trip in _____ seconds.
_____ **5.** If a motor draws 200% of rated current, the overloads will trip in _____ seconds.

◯ Activity 6-5. Determining Overload Heater Size

Determine the heater using Full-Load Current and Heater Selections in Appendix.

_____ **1.** A(n) _____ heater is used with a ¾ HP, 115 V, 1ϕ motor with a size 0 starter.
_____ **2.** A(n) _____ heater is used with a ¾ HP, 230 V, 1ϕ motor with a size 0 starter.
_____ **3.** A(n) _____ heater is used with a 10 HP, 115 V, 1ϕ motor with a size 4 starter.
_____ **4.** A(n) _____ heater is used with a 2 HP, 120 VDC motor with a size 0 starter.
_____ **5.** A(n) _____ heater is used with a 2 HP, 120 VDC motor with a size 1 starter.
_____ **6.** A(n) _____ heater is used with a 2 HP, 120 VDC motor with a size 2 starter.
_____ **7.** A(n) _____ heater is used with an application that has a 50 HP, 230 V, 3ϕ motor.
_____ **8.** A(n) _____ heater is used with a 15 HP, 230 V, 3ϕ motor with a size 2 starter.
_____ **9.** A(n) _____ heater is used with a 20 HP, 460 V, 3ϕ motor with a size 2 starter.
_____ **10.** A(n) _____ heater is used with a ¾ HP, 230 V, 3ϕ motor with a size 0 starter.

◯ Activity 6-6. Contactors

Answer the questions using Control Ratings in Appendix.

_____ **1.** A(n) _____ contactor is used to switch a 1.5 HP, 230 V, 1ϕ (normal duty) motor.
_____ **2.** A(n) _____ contactor is used to switch a 115 V, 1ϕ, 41 A (continuous duty) motor.
_____ **3.** A(n) _____ contactor is used to switch a 27 HP, 460 V, 3ϕ (normal duty) motor.
_____ **4.** A(n) _____ contactor is used to switch a 27 HP, 460 V, 3ϕ (plugging duty) motor.
_____ **5.** A(n) _____ contactor is used to switch a 460 V, 3ϕ, 125 A (continuous duty) motor.
_____ **6.** A(n) _____ contactor is used to switch a 230 V, 50 A tungsten lamp.
_____ **7.** A(n) _____ contactor is used to switch a 115 V, 1ϕ, 50 A heating element.
_____ **8.** A(n) _____ contactor is used to switch a 230 V, 1ϕ, 50 A heating element.
_____ **9.** A(n) _____ contactor is used to switch a 460 V, 3ϕ, 5 kVA (less than 20 times per hour) transformer.
_____ **10.** A(n) _____ contactor is used to switch a 230 V, 10 kVA capacitor bank.

Application — Operate Delay Relays

Operate delay relays are timers that provide a delay period after the relay is energized and before the relay contacts are switched. Operate delay relays are also called ON-delay relays. Operate delay relays are used in safety circuits that require a set time period after a switch is activated and before the load(s) are energized. The delay provides safety by allowing time to review the decision or move out of the way of a process. Manufacturers provide specifications for operate delay relays. **See Models A and B Operate Delay Relays.**

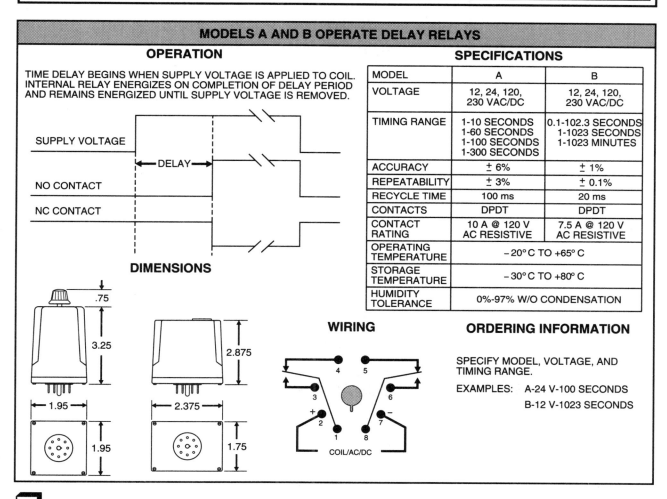

MODELS A AND B OPERATE DELAY RELAYS

OPERATION

TIME DELAY BEGINS WHEN SUPPLY VOLTAGE IS APPLIED TO COIL. INTERNAL RELAY ENERGIZES ON COMPLETION OF DELAY PERIOD AND REMAINS ENERGIZED UNTIL SUPPLY VOLTAGE IS REMOVED.

SUPPLY VOLTAGE

DELAY

NO CONTACT

NC CONTACT

DIMENSIONS

.75
3.25
1.95
1.95

2.875
2.375
1.75

SPECIFICATIONS

MODEL	A	B
VOLTAGE	12, 24, 120, 230 VAC/DC	12, 24, 120, 230 VAC/DC
TIMING RANGE	1-10 SECONDS 1-60 SECONDS 1-100 SECONDS 1-300 SECONDS	0.1-102.3 SECONDS 1-1023 SECONDS 1-1023 MINUTES
ACCURACY	± 6%	± 1%
REPEATABILITY	± 3%	± 0.1%
RECYCLE TIME	100 ms	20 ms
CONTACTS	DPDT	DPDT
CONTACT RATING	10 A @ 120 V AC RESISTIVE	7.5 A @ 120 V AC RESISTIVE
OPERATING TEMPERATURE	−20°C TO +65°C	
STORAGE TEMPERATURE	−30°C TO +80°C	
HUMIDITY TOLERANCE	0%-97% W/O CONDENSATION	

WIRING

4   5
3   6
+
2   7
1   8   −
COIL/AC/DC

ORDERING INFORMATION

SPECIFY MODEL, VOLTAGE, AND TIMING RANGE.

EXAMPLES: A-24 V-100 SECONDS
B-12 V-1023 SECONDS

Application — Release Delay Relays

Release delay relays are timers that provide a delay period after the relay is de-energized. Release delay relays are also called OFF-delay relays. Release delay relays are used in automatic garage door openers that require a time period in which the light is to remain on after the door is closed. Manufacturers provide specifications for release delay relays. **See Models C and D Release Delay Relays.**

MODELS C AND D RELEASE DELAY RELAYS

OPERATION

SUPPLY VOLTAGE IS CONSTANTLY APPLIED TO COIL. INTERNAL RELAYS ENERGIZE WHEN CONTROL SWITCH IS CLOSED. TIMING BEGINS WHEN CONTROL SWITCH IS OPENED. DELAY IS RESET BY RECLOSING CONTROL SWITCH. RELAY DE-ENERGIZES ON COMPLETION OF DELAY PERIOD.

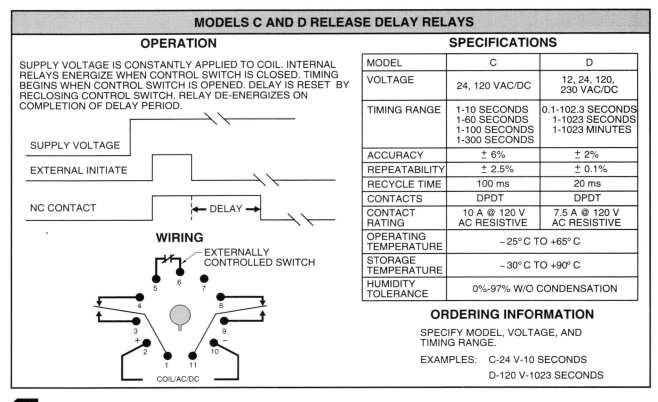

SPECIFICATIONS

MODEL	C	D
VOLTAGE	24, 120 VAC/DC	12, 24, 120, 230 VAC/DC
TIMING RANGE	1-10 SECONDS 1-60 SECONDS 1-100 SECONDS 1-300 SECONDS	0.1-102.3 SECONDS 1-1023 SECONDS 1-1023 MINUTES
ACCURACY	± 6%	± 2%
REPEATABILITY	± 2.5%	± 0.1%
RECYCLE TIME	100 ms	20 ms
CONTACTS	DPDT	DPDT
CONTACT RATING	10 A @ 120 V AC RESISTIVE	7.5 A @ 120 V AC RESISTIVE
OPERATING TEMPERATURE	– 25° C TO +65° C	
STORAGE TEMPERATURE	– 30° C TO +90° C	
HUMIDITY TOLERANCE	0%-97% W/O CONDENSATION	

ORDERING INFORMATION

SPECIFY MODEL, VOLTAGE, AND TIMING RANGE.

EXAMPLES: C-24 V-10 SECONDS

D-120 V-1023 SECONDS

Application — Single Shot Relays

Single shot relays are timers that switch the relay contacts for a set time period after the relay is energized. Single shot relays are also called pulse relays. Single shot relays are used in game machines to provide a time period in which the game can be played after the input is activated. Manufacturers provide specifications for single shot relays. **See Models E and F Single Shot Relays.**

MODELS E AND F SINGLE SHOT RELAYS

OPERATION

INTERNAL RELAY ENERGIZES IMMEDIATELY ON APPLICATION OF SUPPLY VOLTAGE. INTERNAL RELAY DE-ENERGIZES AFTER COMPLETION OF DELAY PERIOD. SUPPLY VOLTAGE MUST BE REMOVED TO RESET TIMER.

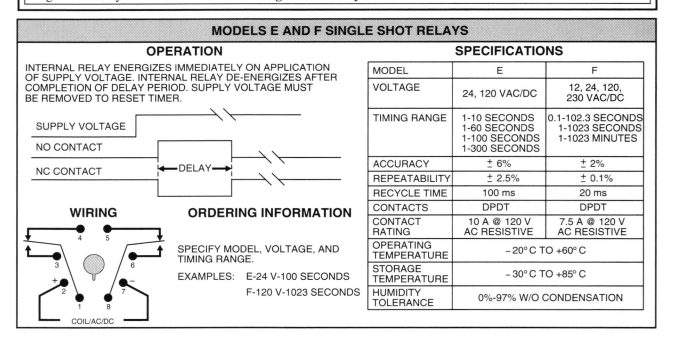

ORDERING INFORMATION

SPECIFY MODEL, VOLTAGE, AND TIMING RANGE.

EXAMPLES: E-24 V-100 SECONDS

F-120 V-1023 SECONDS

SPECIFICATIONS

MODEL	E	F
VOLTAGE	24, 120 VAC/DC	12, 24, 120, 230 VAC/DC
TIMING RANGE	1-10 SECONDS 1-60 SECONDS 1-100 SECONDS 1-300 SECONDS	0.1-102.3 SECONDS 1-1023 SECONDS 1-1023 MINUTES
ACCURACY	± 6%	± 2%
REPEATABILITY	± 2.5%	± 0.1%
RECYCLE TIME	100 ms	20 ms
CONTACTS	DPDT	DPDT
CONTACT RATING	10 A @ 120 V AC RESISTIVE	7.5 A @ 120 V AC RESISTIVE
OPERATING TEMPERATURE	– 20° C TO +60° C	
STORAGE TEMPERATURE	– 30° C TO +85° C	
HUMIDITY TOLERANCE	0%-97% W/O CONDENSATION	

◻ Application — Recycle Timer

Recycle timers are timers that provide an ON/OFF movement of the timer contacts after the timer is energized. Recycle timers are used in automatic lubrication control of motors and machines. Automatic lubrication is required to protect moving parts of motors and machines. A recycle timer controls the period between lubrication. Manufacturers provide specifications for recycle timers. **See Models G and H Recycle Timers.**

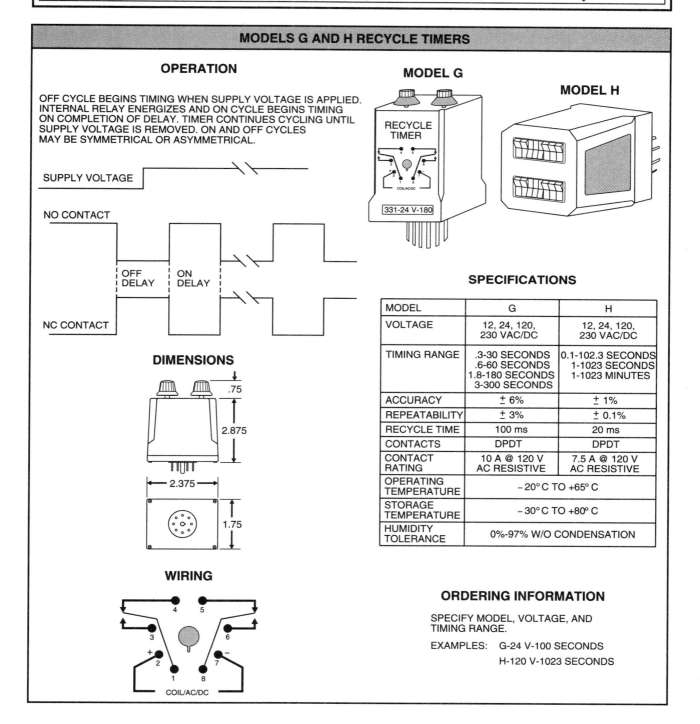

MODELS G AND H RECYCLE TIMERS

OPERATION

OFF CYCLE BEGINS TIMING WHEN SUPPLY VOLTAGE IS APPLIED. INTERNAL RELAY ENERGIZES AND ON CYCLE BEGINS TIMING ON COMPLETION OF DELAY. TIMER CONTINUES CYCLING UNTIL SUPPLY VOLTAGE IS REMOVED. ON AND OFF CYCLES MAY BE SYMMETRICAL OR ASYMMETRICAL.

SUPPLY VOLTAGE

NO CONTACT

OFF DELAY   ON DELAY

NC CONTACT

DIMENSIONS

.75

2.875

2.375

1.75

WIRING

COIL/AC/DC

MODEL G

RECYCLE TIMER

331-24 V-180

MODEL H

SPECIFICATIONS

MODEL	G	H
VOLTAGE	12, 24, 120, 230 VAC/DC	12, 24, 120, 230 VAC/DC
TIMING RANGE	.3-30 SECONDS .6-60 SECONDS 1.8-180 SECONDS 3-300 SECONDS	0.1-102.3 SECONDS 1-1023 SECONDS 1-1023 MINUTES
ACCURACY	± 6%	± 1%
REPEATABILITY	± 3%	± 0.1%
RECYCLE TIME	100 ms	20 ms
CONTACTS	DPDT	DPDT
CONTACT RATING	10 A @ 120 V AC RESISTIVE	7.5 A @ 120 V AC RESISTIVE
OPERATING TEMPERATURE	− 20° C TO +65° C	
STORAGE TEMPERATURE	− 30° C TO +80° C	
HUMIDITY TOLERANCE	0%-97% W/O CONDENSATION	

ORDERING INFORMATION

SPECIFY MODEL, VOLTAGE, AND TIMING RANGE.

EXAMPLES: G-24 V-100 SECONDS

H-120 V-1023 SECONDS

Application — Combination Timing Logic

Circuits often require multiple timing functions in the same circuit. For example, two different timers are used to provide surge and backspin protection when starting and stopping a pump motor. The operate delay relay prevents starting surges from causing the pump to prematurely drop out during starting. This happens because starting surges cause the pressure switch to cycle open and closed. The release delay relay prevents a backspin from immediately restarting the pump once the pump is stopped. This happens because the backspin caused by the pump turnoff could falsely activate the pressure switch. **See Surge and Backspin Protection.**

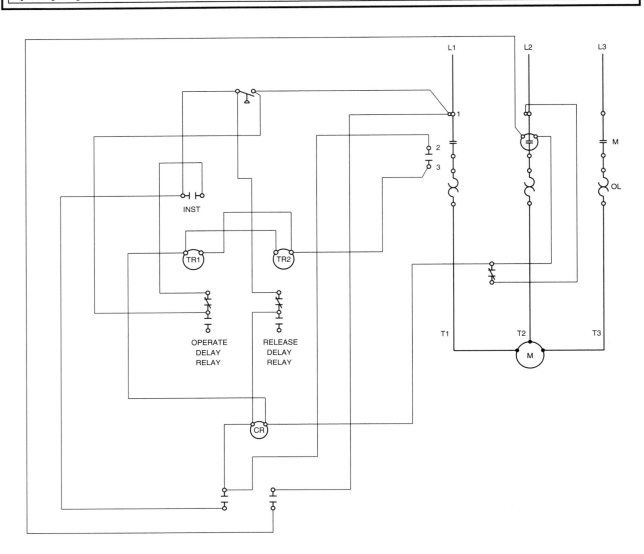

SURGE AND BACKSPIN PROTECTION

ACTIVITIES

Name _____ Date _____

○ Activity 7-1. Operate Delay Relays

A current monitor can be used to detect blade wear. As the blade wears, the drive motor draws more current. When the current reaches a set limit, the current monitor activates the contacts. The current monitor uses a current transformer to detect the amount of motor current. By using different current transformers, currents from .1 A to 500 A can be detected. **See Saw Blade Wear.**

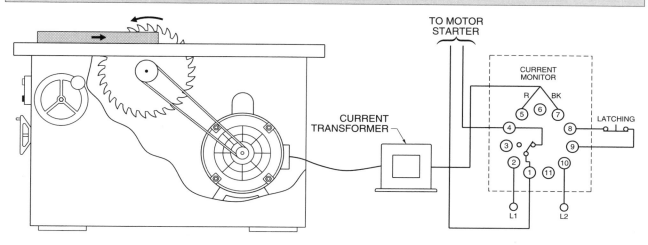

1. Draw the line diagram of Saw Blade Control Circuit using a standard start/stop pushbutton station. Use a magnetic motor starter to control the motor and a start/stop pushbutton station to control the motor starter. Use the motor starter overload contact for overload protection. Do not include the current monitor in the line diagram.

2. In Saw Blade Control Circuit, the motor starter overload contact stops the motor when a sustained overload occurs. The additional current drawn as the saw blade wears is not detected by the motor starter overload. To detect the additional current, a current monitor is used. A current monitor is set to detect a current higher than normal, but lower than a sustained overload.

The current monitor has no built-in time delay. Because all motors draw higher-than-normal current when starting, a timer is used with a current monitor in motor circuits. The timer prevents the current monitor from de-energizing the motor starter during normal motor startup.

Redraw the line diagram of Saw Blade Control Circuit and add a Model A timer and the current monitor. Add the current monitor so the motor starter is de-energized when the current monitor detects blade wear. Add the timer to prevent the current monitor contacts from de-energizing the starter during normal motor startup. Mark all terminal connections from the current monitor and Model A timer. The current monitor is set for 4 A and the overload contacts are set for 6 A. Mark the timer setting for 5 seconds.

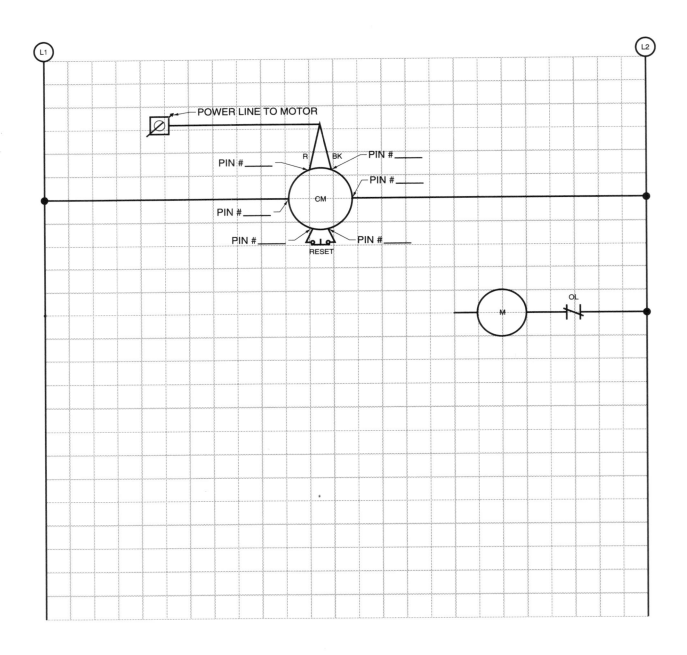

3. In Saw Blade Control Circuit with current monitor, the motor is de-energized if a sustained overload (overload contact) or blade wear (current monitor) occurs. When the overloads de-energize the motor, the motor can be restarted because the overload contact automatically resets. When the current monitor de-energizes the motor, the motor can only be restarted after the reset pushbutton connected to the current monitor is pressed. Redraw the saw blade control circuit with current monitor and add a light that energizes when the current monitor de-energizes the motor. The light indicates the blade is worn and the current monitor stopped the motor. Use a Model G recycle timer to flash the light ON and OFF. Mark all terminal connections.

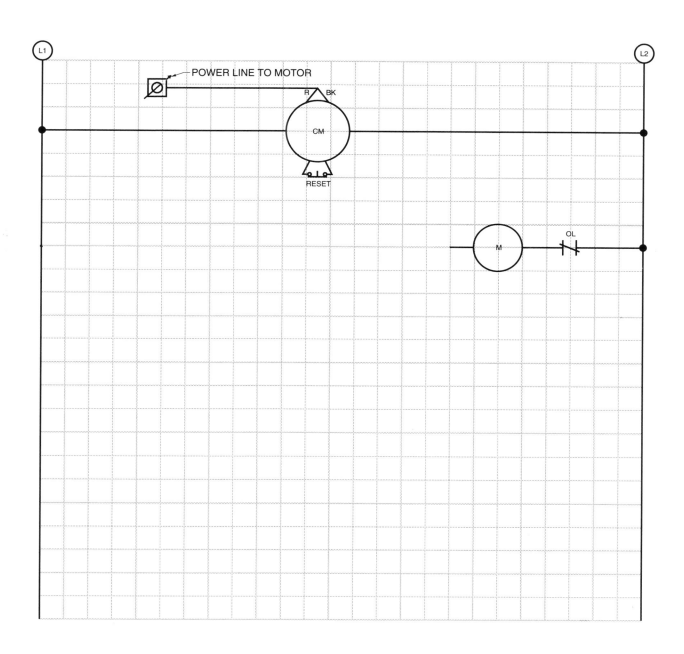

Activity 7-2. Release Delay Relay — Sequential Control

1. Draw the line diagram of the wiring diagram.

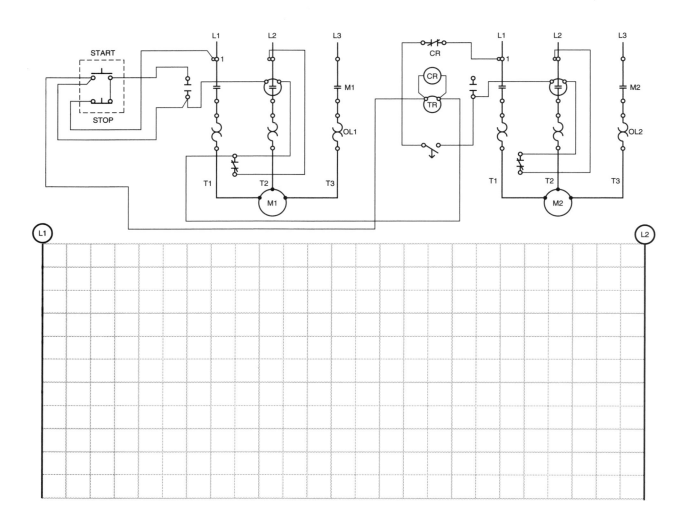

2. Answer the questions using the information and Sequential Control Circuit. Motor Starter 1 controls a pump motor. Motor Starter 2 controls a fan motor. The timer is set for 5 minutes.

_____ **A.** Motor _____ starts first after the start button is pressed.

_____ **B.** Does the other motor start after 5 minutes?

_____ **C.** Motor _____ starts after the stop button is pressed.

_____ **D.** Do both motors ever run at the same time?

_____ **E.** Do both motors ever stop at the same time?

◯ Activity 7-3. Single Shot Relays

Answer the questions using Models E and F Single Shot Relays on page 60.

_____ 1. A Model _____ timer has a period of 1 second to 60 seconds and coil voltage of 120 VAC.

_____ 2. A Model _____ timer has a period of 1 second to 1023 seconds and coil voltage of 24 VAC.

_____ 3. What are the pin numbers of the timer coil?

_____ 4. What are the pin numbers of a normally open timer contact?

_____ 5. What are the pin numbers of a normally closed timer contact?

_____ 6. Which of the two timers, Model E or Model F, provides the most accurate time setting?

_____ 7. The maximum current that can be switched with a Model E timer is _____ A.

_____ 8. The maximum current that can be switched with a Model F timer is _____ A.

_____ 9. A(n) _____ timer has the greatest repeatable accuracy.

_____ 10. After a Model E timer is reset, _____ seconds are required to recycle.

_____ 11. A Model _____ timer provides the longest timing function.

◯ Activity 7-4. Recycle Timer

Draw the line diagram of a control circuit using the Model G recycle timer so the timer starts running when a two-position selector switch is placed in the automatic position. The timer is set for a 4 hour OFF-delay period. After the 4 hours, the timer contacts energize a solenoid. The solenoid controls the lubrication flow. After a short ON-time period, the timer turns OFF the solenoid for another 4 hours. When the selector switch is placed in the OFF position, the timer stops running. Mark the timer pin numbers on the circuit.

Note: The recycle timer energizes the solenoid for a minimum of one minute when the 1 minute to 1023 minute time range is used. If a shorter solenoid-energized time period is required, the recycle timer can control a single shot timer. The single shot timer controls the lubrication solenoid.

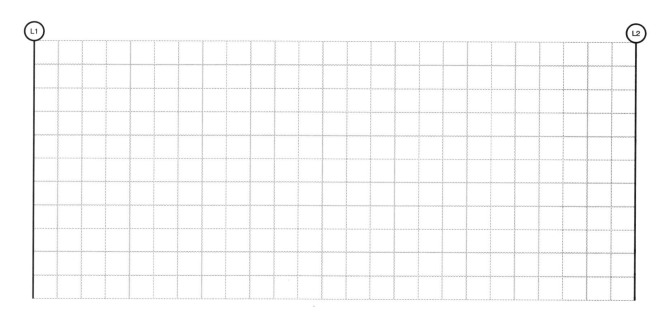

Activity 7-5. Combination Timing Logic

Draw the line diagram for the circuit using the Model A operate delay and Model C release delay relays. Timers are combined for sensitive heat control. Timer 1 is set for 5 minutes and Timer 2 is set for 10 minutes. Timer 1 ensures that the heater is ON for at least 5 minutes, even if the temperature switch opens before 5 minutes. Timer 2 prevents the heater from turning back ON for at least 10 minutes, even if the temperature switch closes. This combination ensures an even and economical heat supply. Draw the line diagram of the circuit.

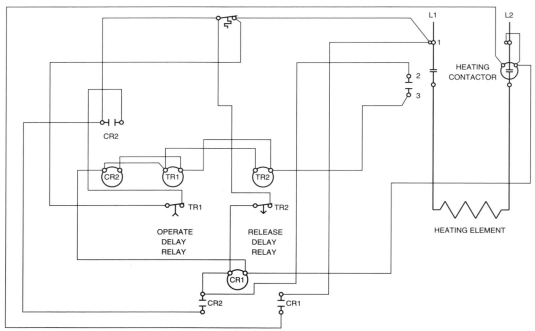

APPLICATIONS

▢ Application — Alternating Motor Control

Using dual motors in an application prevents downtime and reduced production. The two main advantages of using dual motors is that one motor can do the work if the other motor is down, and both motors can be energized if extra work is required.

The three control options possible with dual-motor circuits are:

1. One motor (main motor) is used each time work is required. The other motor (back-up motor) is used when a problem occurs with the main motor. The disadvantage is that one motor is always working and the other is always idle, which is not good for either motor.

2. One motor is used for a time period (day or week) and the other motor remains idle. After the time period, the motor functions are switched. The disadvantage is that a rotation schedule is required.

3. The motors are alternated from rest to work each time work is required. The motor alteration is accomplished through an alternating (flip-flop) relay. The advantage is that both motors are ON approximately the same amount of time over the long run. This is the best method for most dual-motor applications. **See Flip-flop Relay.**

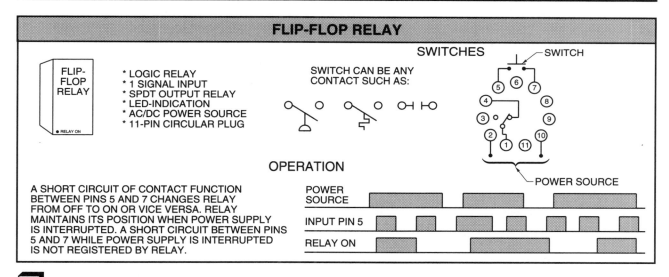

FLIP-FLOP RELAY

FLIP-FLOP RELAY

* LOGIC RELAY
* 1 SIGNAL INPUT
* SPDT OUTPUT RELAY
* LED-INDICATION
* AC/DC POWER SOURCE
* 11-PIN CIRCULAR PLUG

● RELAY ON

SWITCHES

SWITCH CAN BE ANY CONTACT SUCH AS:

OPERATION

A SHORT CIRCUIT OF CONTACT FUNCTION BETWEEN PINS 5 AND 7 CHANGES RELAY FROM OFF TO ON OR VICE VERSA. RELAY MAINTAINS ITS POSITION WHEN POWER SUPPLY IS INTERRUPTED. A SHORT CIRCUIT BETWEEN PINS 5 AND 7 WHILE POWER SUPPLY IS INTERRUPTED IS NOT REGISTERED BY RELAY.

POWER SOURCE

INPUT PIN 5

RELAY ON

▢ Application — Level Control

Level controls are relays that maintain a set level in a tank. A level control keeps a tank full (charging) or empty (discharging). *Charging* is the process of filling or keeping a tank full. Charging a tank is accomplished by a pump motor or a solenoid valve. A solenoid valve is used in applications where the tank to be charged is located under the supply tank or where the product to be placed in the tank is under pressure. A pump motor is used in applications where the supply tank is level with or below the tank to be charged.

Discharging is the process of emptying a tank. Like charging, discharging a tank can be accomplished by a pump motor or a solenoid valve. Whether a pump motor or a valve is used for discharging depends on the viscosity of the product, the location of the receiving tank, and the distance the product must be moved. Most tanks are discharged by a pump motor.

To control the charging or discharging of a tank, a level-control relay is used. The level-control relay controls the pump motor or valve. Level probes that detect the product inside the tank are connected to the relay. **See Level-control Relay.**

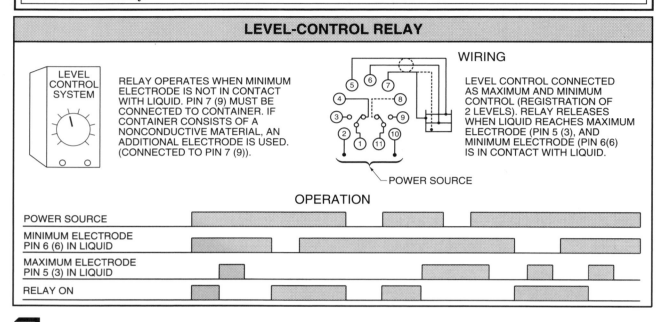

LEVEL-CONTROL RELAY

LEVEL CONTROL SYSTEM

RELAY OPERATES WHEN MINIMUM ELECTRODE IS NOT IN CONTACT WITH LIQUID. PIN 7 (9) MUST BE CONNECTED TO CONTAINER. IF CONTAINER CONSISTS OF A NONCONDUCTIVE MATERIAL, AN ADDITIONAL ELECTRODE IS USED. (CONNECTED TO PIN 7 (9)).

WIRING

LEVEL CONTROL CONNECTED AS MAXIMUM AND MINIMUM CONTROL (REGISTRATION OF 2 LEVELS). RELAY RELEASES WHEN LIQUID REACHES MAXIMUM ELECTRODE (PIN 5 (3), AND MINIMUM ELECTRODE (PIN 6(6) IS IN CONTACT WITH LIQUID.

POWER SOURCE

OPERATION

POWER SOURCE

MINIMUM ELECTRODE PIN 6 (6) IN LIQUID

MAXIMUM ELECTRODE PIN 5 (3) IN LIQUID

RELAY ON

Application — Temperature Control

Temperature controls are temperature-sensitive relays that maintain the proper temperature in an application. Depending on the contacts (NO or NC), the temperature switch is used for heating or cooling applications. With some temperature controls, cooling or heating control is determined by a switch or interconnecting pins. **See Temperature-control Relay.**

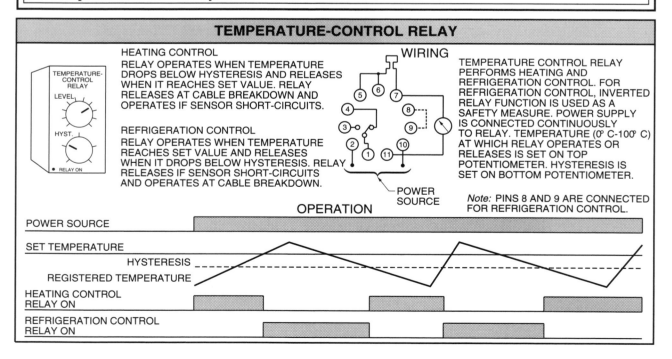

TEMPERATURE-CONTROL RELAY

TEMPERATURE-CONTROL RELAY

LEVEL

HYST.

RELAY ON

HEATING CONTROL
RELAY OPERATES WHEN TEMPERATURE DROPS BELOW HYSTERESIS AND RELEASES WHEN IT REACHES SET VALUE. RELAY RELEASES AT CABLE BREAKDOWN AND OPERATES IF SENSOR SHORT-CIRCUITS.

REFRIGERATION CONTROL
RELAY OPERATES WHEN TEMPERATURE REACHES SET VALUE AND RELEASES WHEN IT DROPS BELOW HYSTERESIS. RELAY RELEASES IF SENSOR SHORT-CIRCUITS AND OPERATES AT CABLE BREAKDOWN.

WIRING

POWER SOURCE

TEMPERATURE CONTROL RELAY PERFORMS HEATING AND REFRIGERATION CONTROL. FOR REFRIGERATION CONTROL, INVERTED RELAY FUNCTION IS USED AS A SAFETY MEASURE. POWER SUPPLY IS CONNECTED CONTINUOUSLY TO RELAY. TEMPERATURE (0° C-100° C) AT WHICH RELAY OPERATES OR RELEASES IS SET ON TOP POTENTIOMETER. HYSTERESIS IS SET ON BOTTOM POTENTIOMETER.

Note: PINS 8 AND 9 ARE CONNECTED FOR REFRIGERATION CONTROL.

OPERATION

POWER SOURCE

SET TEMPERATURE

HYSTERESIS

REGISTERED TEMPERATURE

HEATING CONTROL RELAY ON

REFRIGERATION CONTROL RELAY ON

Name _____   Date _____

Activity 8-1.   Alternating Motor Control

Draw the line diagram of the control circuit using Dual Pump Application and Flip-flop Relay. Include a magnetic motor starter (M1) that controls pump motor 1 and a magnetic motor starter (M2) that controls pump motor 2. Include a liquid level switch that energizes a control relay (CR-1), and a flip-flop relay that is controlled by the control relay (CR-1) each time the liquid level switch energizes the relay.

Note: One set of relay contacts is connected across pins 5 and 7 of the flip-flop relay. Connect M1 through one set of contacts (1 and 3) from the flip-flop relay, and M2 through the other set of contacts (1 and 4) from the flip-flop relay. Connect the flip-flop relay pin 1 in series with a second set of normally open contacts from the control relay (CR-1). Connect the second set of control relay contacts to power line 1. No power is applied to the flip-flop contacts (and the pump motors) until the liquid level switch energizes the control relay.

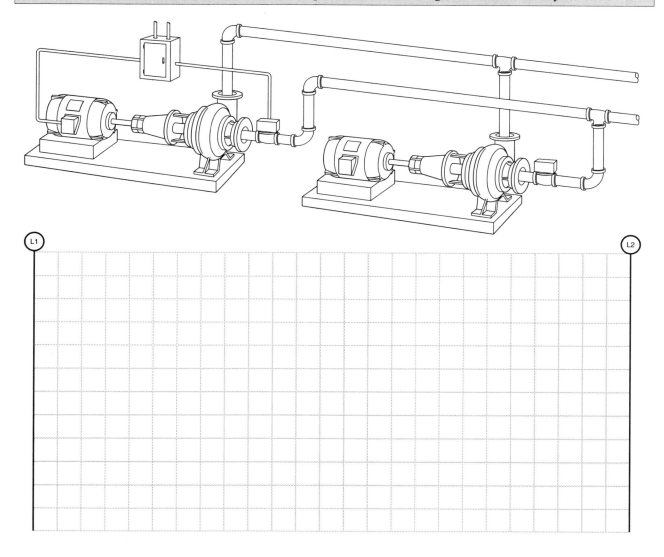

◯ Activity 8-2. Level Control

Draw the line diagram of the control circuit for the tank application and the level-control relay. Include a solenoid valve that opens a valve that fills the tank. The solenoid valve is controlled by the level control relay. Include a magnetic motor starter to control a pump motor that empties the tank. The magnetic motor starter is controlled by a press-to-empty pushbutton. The tank can be manually emptied using this pushbutton. Include a selector switch to place the circuit in a hand/OFF/auto position. In the auto position, the level relay automatically keeps the tank full. In the hand position, the fill solenoid is energized regardless of the level in the tank. In the OFF position, no filling takes place.

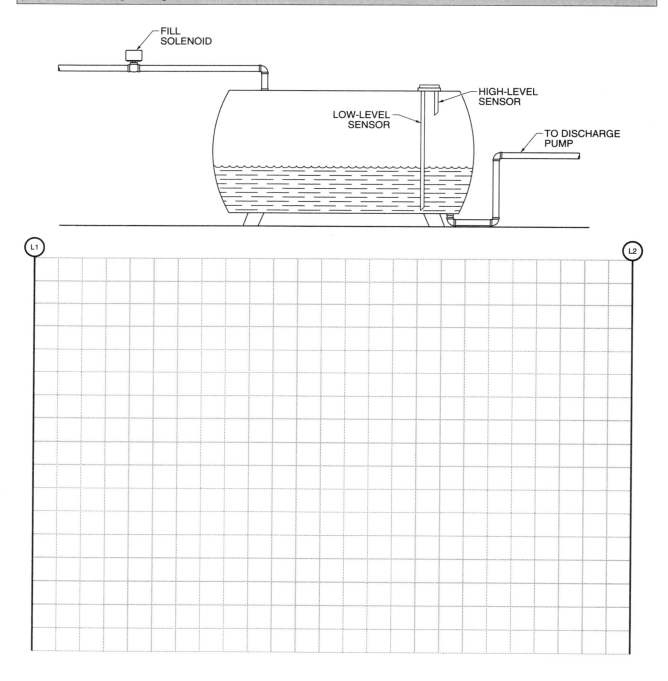

⬤ Activity 8-3. Temperature Control

1. Draw the line diagram of the control circuit for the tank application and the temperature-control relay. Use a contactor to control a heating element. The temperature control controls the heating contactor. A selector switch places the circuit in a hand/OFF/auto position. In the auto position, the temperature relay keeps the product at set temperature. In the hand position, the heating contactor is energized regardless of the temperature. In the OFF position, no heating takes place. Mark each pin number used on the temperature relay.

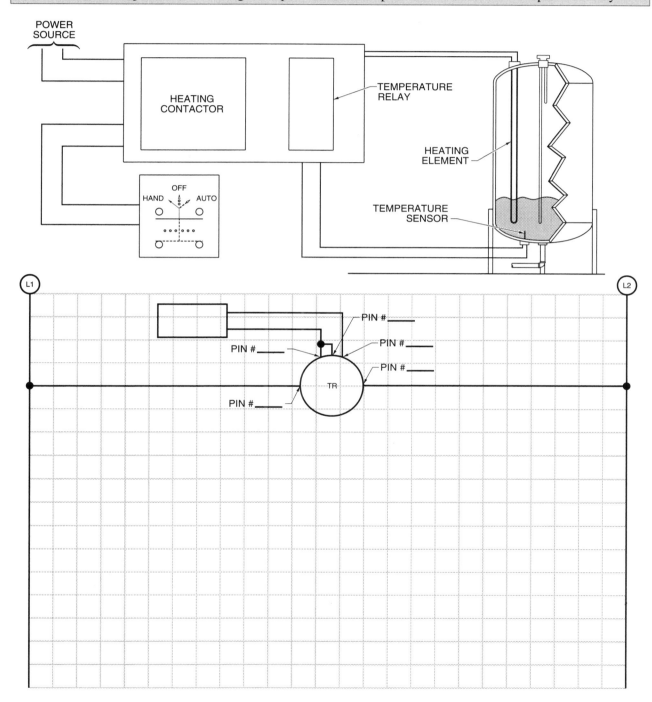

2. Combine the level- and temperature-control circuits so no heating takes place unless the tank is full of product.

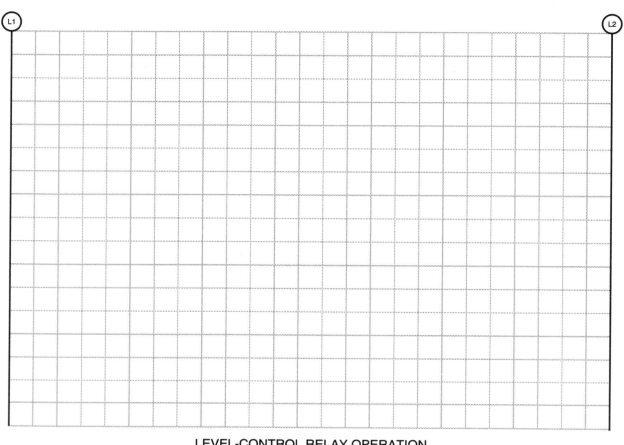

LEVEL-CONTROL RELAY OPERATION

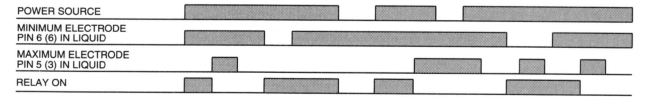

POWER SOURCE	
MINIMUM ELECTRODE PIN 6 (6) IN LIQUID	
MAXIMUM ELECTRODE PIN 5 (3) IN LIQUID	
RELAY ON	

TEMPERATURE-CONTROL RELAY OPERATION

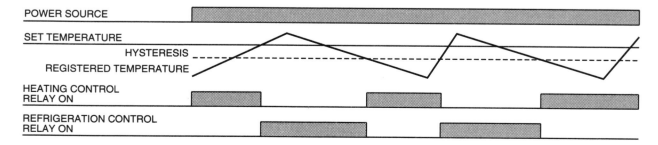

POWER SOURCE	
SET TEMPERATURE	
HYSTERESIS	
REGISTERED TEMPERATURE	
HEATING CONTROL RELAY ON	
REFRIGERATION CONTROL RELAY ON	

Application — Reversing Motors

Motors are designed to rotate in the clockwise or counterclockwise direction. Clockwise (FWD) or counterclockwise (REV) rotation is determined by viewing the front of the motor. Normally, the front of a motor is the end opposite the shaft, and the back of a motor is the end with the shaft.

Any motor that is designed to rotate in both directions can be reversed. Reversing a motor is accomplished by interchanging motor leads. The motor leads are connected to a drum switch, manual reversing starter, or magnetic reversing starter. **See Reversing Methods.**

Reversing Drum Switch

A *reversing drum switch* is a three-position manual switch that has six terminal connections. The motor leads are connected to the terminal connections. A drum switch interchanges the motor leads and does not provide overload protection. Drum switches are normally used with fractional horsepower motors that control machine tools, cranes, and forklifts.

Manual Reversing Starter

A *manual reversing starter* is a switch that includes overload protection. Overload protection protects the motor when running and automatically disconnects an overloaded motor. To prevent a short circuit and damage to the motor, the starter provides mechanical interlocking. *Mechanical interlocking* prevents the forward and reverse contacts from being energized at the same time. Manual reversing starters are normally used with motors less than 2 HP.

Magnetic Reversing Starter

A *magnetic reversing starter* is a switch that uses starting coils to control the position of the contacts. By using starting coils, the starter can be remotely controlled. The forward coil energizes the forward set of contacts, and the reverse coil energizes the reversing set of contacts. A magnetic reversing starter provides mechanical interlocking. Magnetic reversing starters are used with all sizes of motors and are the most common starter used in industrial applications. The magnetic reversing starter includes overload protection.

Note: Wire numbering systems vary by motor manufacturer. Wire numbers may be given on the nameplate, wires, or near terminal connections.

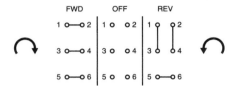

REVERSING DRUM SWITCH

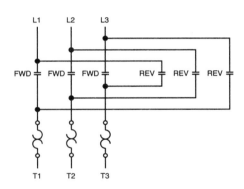

MANUAL REVERSING STARTER

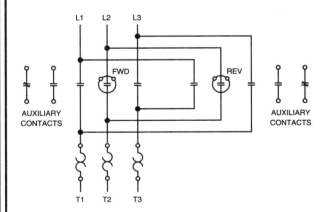

MAGNETIC REVERSING STARTER

Application — Reversing 3ϕ Motors

Three-phase motors are reversed by interchanging two of the three main power lines of the motor. The industry standard is to interchange lines one (L1) and three (L3). The motor is connected L1 to T1, L2 to T2, and L3 to T3 for one direction of rotation, and L3 to T1, L2 to T2, and L1 to T3 for the opposite direction of rotation. **See 3ϕ Motor Wiring Diagrams.**

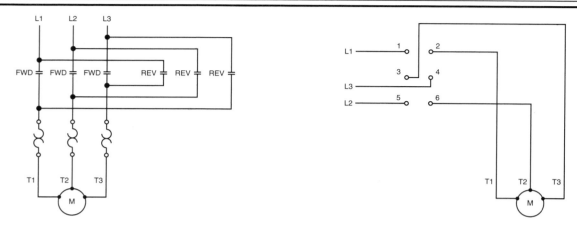

3ϕ MOTOR REVERSING STARTER

3ϕ MOTOR REVERSING DRUM SWITCH

Application — Reversing 1ϕ Motors

A 1ϕ motor requires a starting winding for starting. The starting winding gives the motor starting torque and determines the direction of motor rotation. A 1ϕ motor is reversed by interchanging the starting winding leads. Reversing the starting windings is accomplished through the motor starter. The motor starter is used to interchange the starting winding leads and to disconnect the running windings from power. **See 1ϕ Motor Wiring Diagrams.**

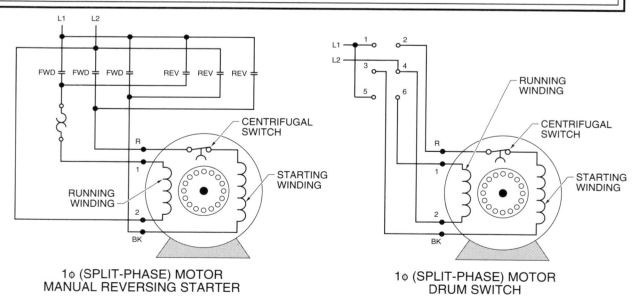

1ϕ (SPLIT-PHASE) MOTOR
MANUAL REVERSING STARTER

1ϕ (SPLIT-PHASE) MOTOR
DRUM SWITCH

Capacitor-start Motors

Capacitor-start motors have a capacitor added in series with the starting winding of a 1φ motor. The capacitor is located outside the motor housing. The capacitor provides additional starting torque by adjusting the phase angle between the voltage and current in the motor windings. A centrifugal switch is used to disconnect the capacitor and the starting windings from power when the motor reaches a set speed. To reverse the rotation of a capacitor-start motor, the starting winding leads are interchanged. **See Capacitor-start Motor.**

Dual-capacitor Motors

Two capacitors give 1φ motors more starting and running torque. One capacitor is sized for starting, and the other capacitor is sized for running. A larger value capacitor is used for starting, and a smaller value capacitor is used for running. The starting capacitor is connected in series with the starting winding, and the running capacitor is connected in series with the starting winding after the centrifugal switch opens. To reverse the rotation of a dual-capacitor motor, the starting winding leads are interchanged. **See Dual-capacitor Motor.**

Application — Reversing Dual-voltage Motors

Most industrial motors are dual-voltage motors. *Dual-voltage motors* are motors that are rated at two different voltages. A dual-voltage, 3φ motor is normally rated at 240/480 V. A dual-voltage, 1φ motor is normally rated at 120/240 V. The higher voltage is used where possible because the power is the same as the lower voltage but one-half of the current is drawn. With less current drawn, wire size and installation cost is reduced.

To allow a motor to be connected to two different voltages (120/240 V), the main running winding is divided into two parts. The two running windings are connected in series for the high voltage and in parallel for the low voltage. The starting winding is connected across one of the running windings. All motor windings receive nearly the same voltage when wired for low- or high-voltage operation. **See Dual-voltage Motor.**

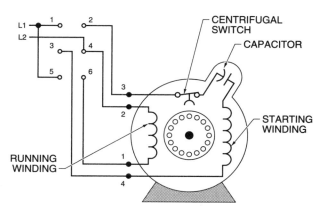

CAPACITOR-START MOTOR

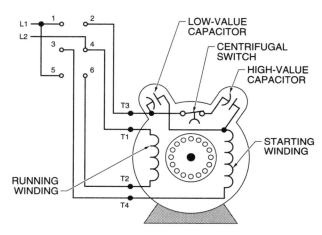

DUAL-CAPACITOR MOTOR

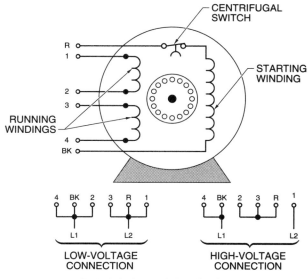

DUAL-VOLTAGE MOTOR

Application — Reversing DC Motors

The rotation of a DC series, shunt, or compound motor depends on the direction of current flow in the field circuit and the armature circuit. To reverse the rotation, the current direction in the field or the armature is reversed. Normally, the current through the armature is reversed. **See DC Motors.** Standard abbreviations are used with DC motors: A1, A2 = armature; F1, F2 = shunt field; S1, S2 = series field.

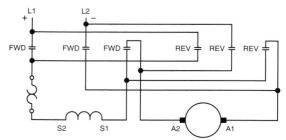

DC SERIES MOTOR REVERSING STARTER

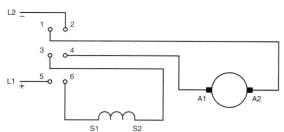

DC SERIES MOTOR DRUM SWITCH

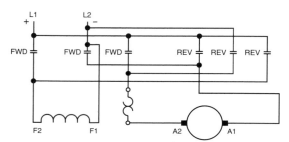

DC SHUNT MOTOR REVERSING STARTER

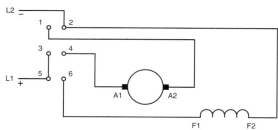

DC SHUNT MOTOR DRUM SWITCH

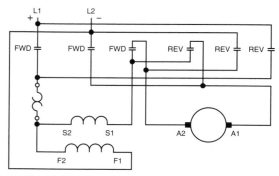

DC COMPOUND MOTOR REVERSING STARTER

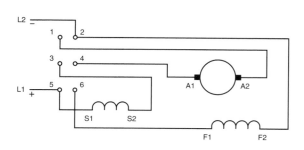

DC COMPOUND MOTOR DRUM SWITCH

ACTIVITIES

Name _____ Date _____

◯ **Activity 9-1. Reversing Motors**

1. State the connections for a 3φ motor and a motor starter.

Forward
_____ **A.** _____ to T1
_____ **B.** _____ to T2
_____ **C.** _____ to T3

Reverse
_____ **D.** _____ to T1
_____ **E.** _____ to T2
_____ **F.** _____ to T3

2. State the connections for a 3φ motor and a drum switch.

Forward
_____ **A.** _____ to T1
_____ **B.** _____ to T2
_____ **C.** _____ to T3

Reverse
_____ **D.** _____ to T1
_____ **E.** _____ to T2
_____ **F.** _____ to T3

3. State the connections for a 1φ (split-phase) motor and a drum switch.

Forward
_____ **A.** _____ to red
_____ **B.** _____ to black
_____ **C.** _____ to 1

Reverse
_____ **D.** _____ to red
_____ **E.** _____ to black
_____ **F.** _____ to 1

4. State the connections for a 1φ (split-phase) motor and a motor starter.

Forward
_____ **A.** _____ to 1
_____ **B.** _____ to red
_____ **C.** _____ to black

Reverse
_____ **D.** _____ to 1
_____ **E.** _____ to red
_____ **F.** _____ to black

5. State the connections for a capacitor-start motor and a drum switch.

Forward
_____ **A.** _____ to 3
_____ **B.** _____ to 4
_____ **C.** _____ to 2

Reverse
_____ **D.** _____ to 3
_____ **E.** _____ to 4
_____ **F.** _____ to 2

6. State the connections for a dual-capacitor motor and a drum switch.

Forward

_____ **A.** _____ to T1

_____ **B.** _____ to T4

_____ **C.** _____ to T3

Reverse

_____ **D.** _____ to T1

_____ **E.** _____ to T4

_____ **F.** _____ to T3

7. State the connections for a DC shunt motor and a drum switch.

Forward

_____ **A.** _____ to A1

_____ **B.** _____ to A2

_____ **C.** _____ to F1

Reverse

_____ **D.** _____ to A1

_____ **E.** _____ to A2

_____ **F.** _____ to F1

8. State the connections for a DC series compound motor and a motor starter.

Forward

_____ **A.** _____ to L2/F2

_____ **B.** _____ to S1

_____ **C.** _____ to S2

Reverse

_____ **D.** _____ to L2/F2

_____ **E.** _____ to S1

_____ **F.** _____ to S2

9. Draw the line diagram of Motor Control Circuit. Include the limit switches.

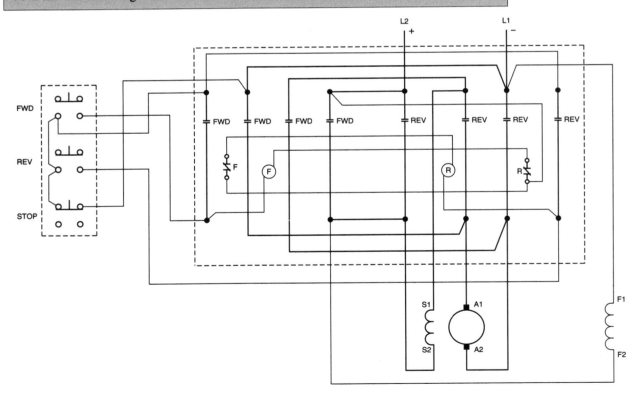

MOTOR CONTROL CIRCUIT

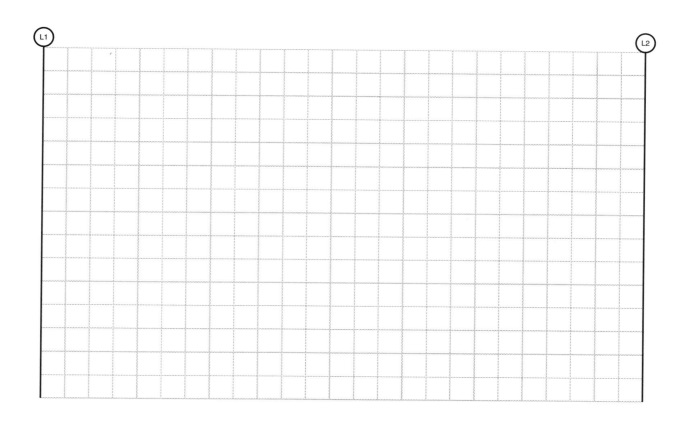

10. Answer the questions using the motor control circuit on page 80.

_____ **A.** If the motor is not running, can the motor be started in either the forward or reverse direction?

_____ **B.** If the motor is running in the forward direction, will pressing the reverse pushbutton energize the reverse starter coil?

_____ **C.** If the motor is running in the reverse direction, will pressing the forward pushbutton change the direction of motor rotation?

_____ **D.** Does the motor starter provide overload protection?

_____ **E.** Does the stop pushbutton have to be pressed to change the direction of rotation when the motor is running?

_____ **F.** A(n) _____ DC motor is changing direction.

Activity 9-2. Reversing 1φ Motors

Wire the motor for forward and reversing.

ROTATION	L1	L2
FWD	1, 8	4, 5
REV	1, 5	4, 8

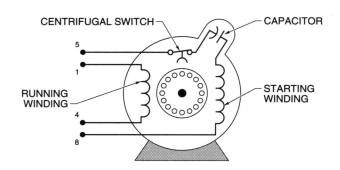

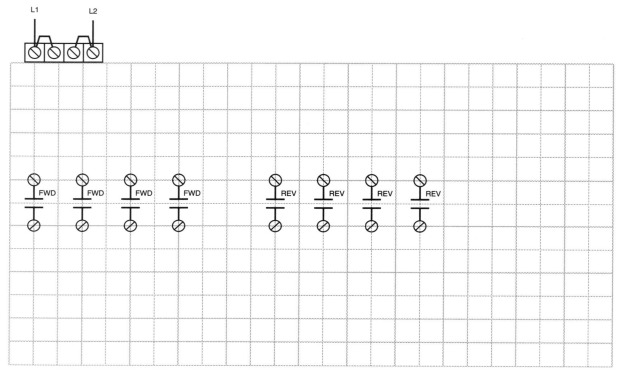

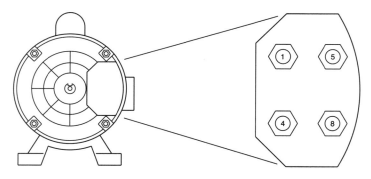

◯ Activity 9-3. Reversing Dual-voltage Motors

1. Wire the motor for forward and reversing using low voltage.

	ROTATION	L1	L2	JOIN
HIGH VOLTAGE	FWD	1	4, 5	2 & 3 & 8
	REV	1	4, 8	2 & 3 & 5
LOW VOLTAGE	FWD	1, 3, 8	2, 4, 5	–
	REV	1, 3, 5	2, 4, 8	–

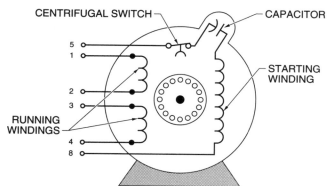

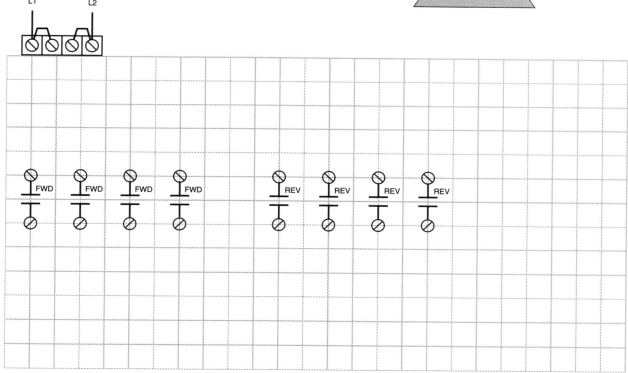

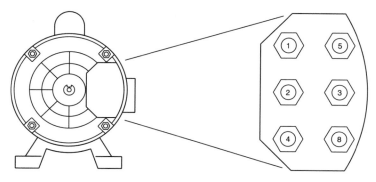

2. Wire the motor for forward and reversing using high voltage.

	ROTATION	L1	L2	JOIN
HIGH VOLTAGE	FWD	1	4, 5	2 & 3 & 8
	REV	1	4, 8	2 & 3 & 5
LOW VOLTAGE	FWD	1, 3, 8	2, 4, 5	–
	REV	1, 3, 5	2, 4, 8	–

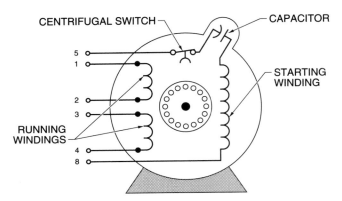

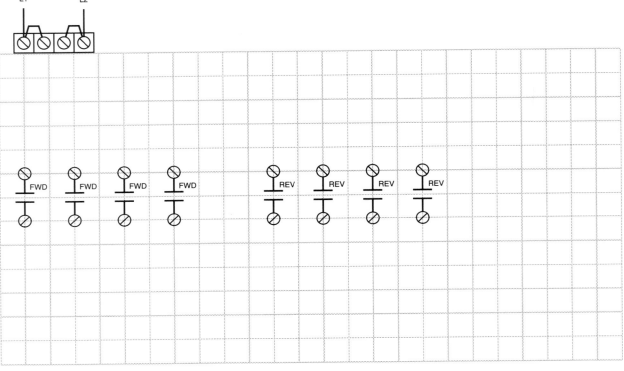

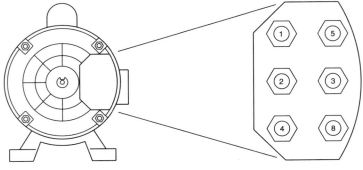

3. Wire the motor for forward and reversing using low voltage.

	ROTATION	L1	L2	JOIN
HIGH VOLTAGE	FWD	1, 8	4, 5	2 & 3, 6 & 7
	REV	1, 5	4, 8	2 & 3, 6 & 7
LOW VOLTAGE	FWD	1, 3, 6, 8	2, 4, 5, 7	–
	REV	1, 3, 5, 7	2, 4, 6, 8	–

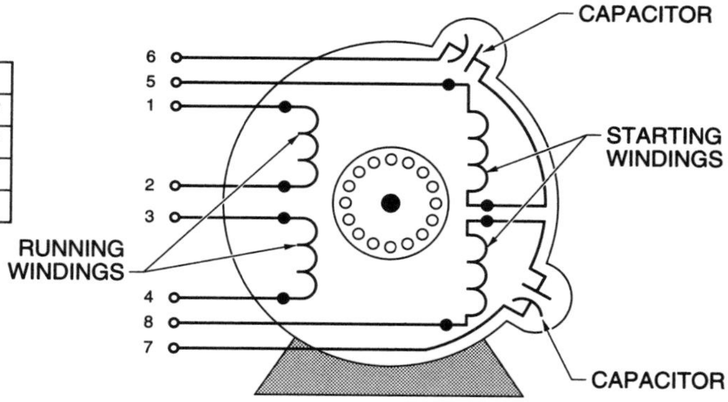

CAPACITOR

STARTING WINDINGS

RUNNING WINDINGS

CAPACITOR

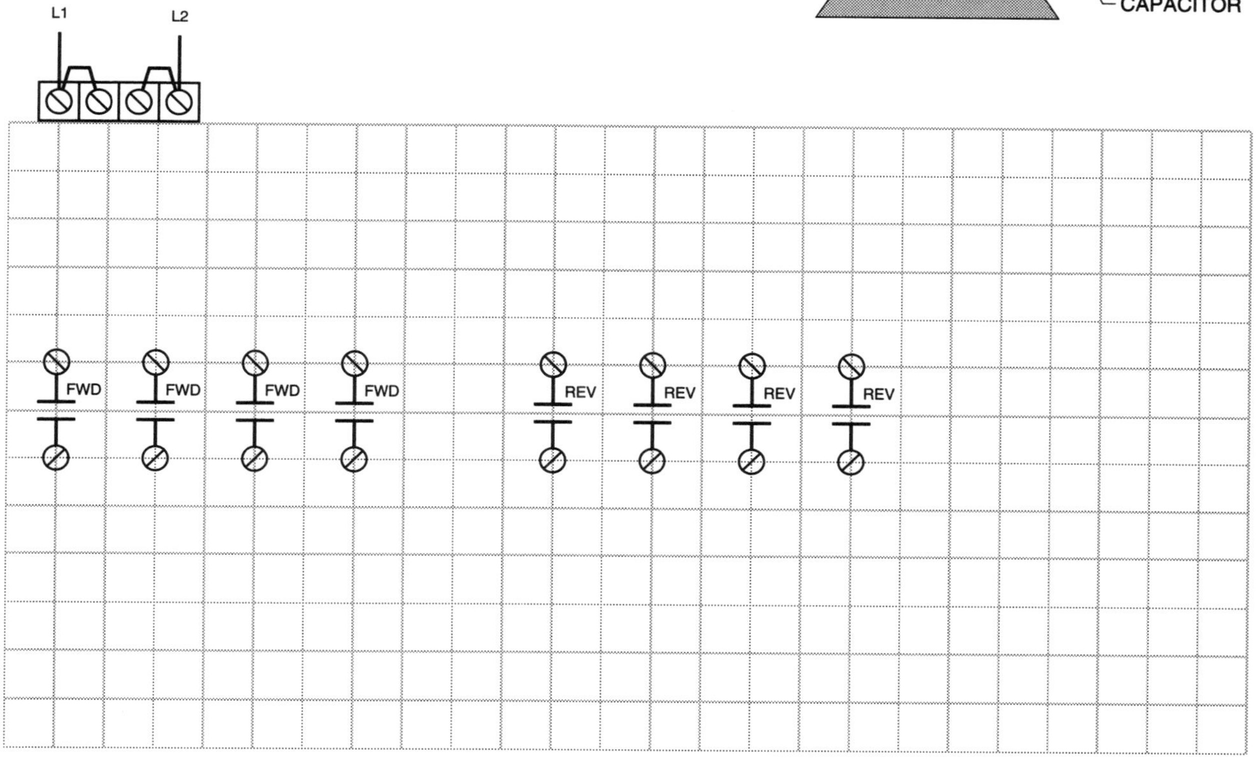

L1   L2

FWD   FWD   FWD   FWD   REV   REV   REV   REV

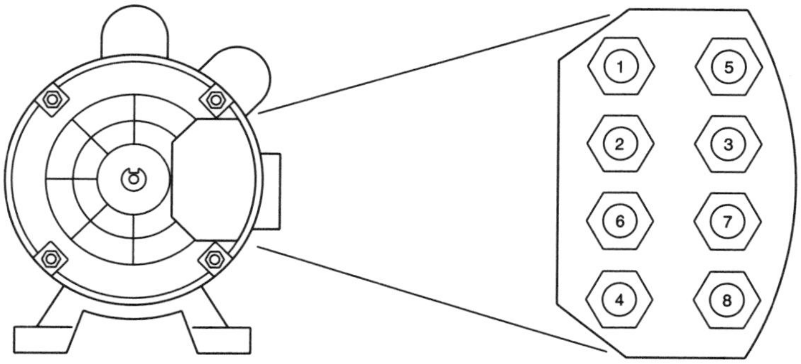

4. Wire the motor for forward and reversing using high voltage.

	ROTATION	L1	L2	JOIN
HIGH VOLTAGE	FWD	1, 8	4, 5	2 & 3, 6 & 7
	REV	1, 5	4, 8	2 & 3, 6 & 7
LOW VOLTAGE	FWD	1, 3, 6, 8	2, 4, 5, 7	–
	REV	1, 3, 5, 7	2, 4, 6, 8	–

Application — Transformer Tap Connection

A transformer with taps is used to compensate for voltage differences in a control circuit. *Taps* are connecting points that are provided along the transformer coil. Taps are usually provided at $2\frac{1}{2}°$ increments along one end of the transformer coil. **See Transformer Nameplate.**

To determine the proper tap connections, apply the procedure:

Step 1. Measure the incoming voltage on the primary side of the transformer using a voltmeter.

Step 2. Determine the secondary voltage. The secondary voltage is the voltage rating of the load(s) that are connected to the transformer.

Step 3. Determine the connections using the manufacturer's nameplate or specifications for the transformer.

Step 4. Shut the power OFF, lock out, and tag the incoming power to the transformer. Make sure the power is OFF by testing the circuit with a voltmeter.

Step 5. Connect the transformer as listed by the manufacturer. Check each connection twice.

Step 6. Turn the power ON.

Step 7. Measure the secondary voltage of the transformer.

Step 8. If the secondary voltage is not correct, repeat Steps 1–6.

TRANSFORMER

DRY TYPE	INDOOR	3ϕ	60 Hz	CLASS AA

			JUMPER CONNECTIONS EACH PHASE	
MODEL # T624A762			**VOLTS**	**TAP**
SERIAL # 68A			503	1
kVA 50	150°C RISE		493	2
HV 480	V LINE-TO-LINE		480	3
LV 208	V LINE-TO-LINE		466	4
LV 120	V LINE-TO-NEUTRAL		456	5
WEIGHT 400	LB		443	6
			433	7
H1, H2, H3 = HIGH SIDE				
X1, X2, X3 = LOW SIDE				

HOMEWOOD, IL   MADE IN USA

Application — Wye and Delta Transformer Connections

Transformers are connected in wye and delta configurations. A *wye configuration* is a transformer connection that has one end of each transformer coil connected together. The remaining end of each coil is connected to the incoming power lines (primary side) or used to supply power to the load(s) (secondary side). A *delta configuration* is a transformer connection that has each transformer coil connected end-to-end to form a closed loop. Each connecting point is connected to the incoming power lines or used to supply power to the load(s). The voltage output and type available for the load(s) is determined by whether a transformer is connected in a wye or delta configuration. **See Wye and Delta Transformer Configurations.**

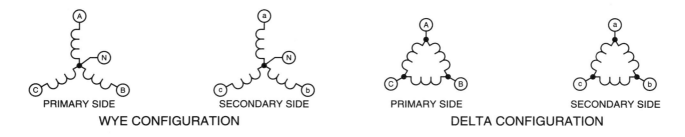

PRIMARY SIDE   SECONDARY SIDE

WYE CONFIGURATION

PRIMARY SIDE   SECONDARY SIDE

DELTA CONFIGURATION

Application — Motor Control Center Assembly

Applications that require several motor starters assembled in a group normally have the motor starters, disconnect switches, control circuit, and components combined in a motor control center. A *motor control center* is a sheet metal enclosure that encloses and protects motor starters, fuses, circuit breakers, overloads, and wiring.

To assemble a motor control center, apply the procedure:

Step 1. Select the installation location. The motor control center is located as close to the application and loads as possible because the center is used when troubleshooting the system. When selecting a location, the floor must be level and the wall or supporting structure should be plumb. **See Installation Location. Caution:** If the control center is not free-standing, conduit should not be used as the only support for the control center.

Step 2. Assemble the frame. Standard motor control centers consist of units that are 20″ wide by 15″ or 20″ deep and 90″ high. The units are assembled then secured together.

The control center frame consists of vertical and horizontal supports. The vertical supports are assembled first. The mounting bolts are loosely tightened until all frame supports are in place, then tightened as recommended by the manufacturer. Back or side panels are left off to allow for wiring the unit. **See Frame Assembly.**

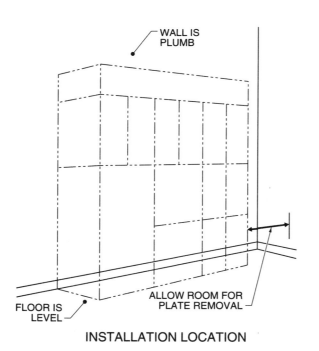

INSTALLATION LOCATION

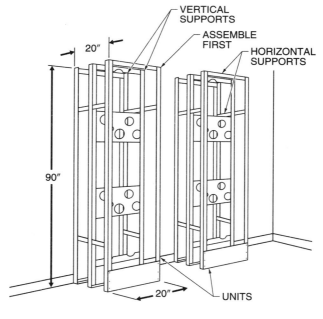

FRAME ASSEMBLY

Step 3. Install the power bus. Power is distributed through the motor control center by the power bus. The power bus consists of a horizontal bus (main) and vertical buses (units). The horizontal bus runs the length of the center, passing through each unit. Each unit has a vertical bus that is connected to the horizontal bus. The vertical bus runs the full height of each unit. The vertical bus supplies power to each motor starter. The horizontal bus is assembled first, and then the vertical buses are assembled. After the vertical buses are in place, bus supports are installed. **See Bus Installation.**

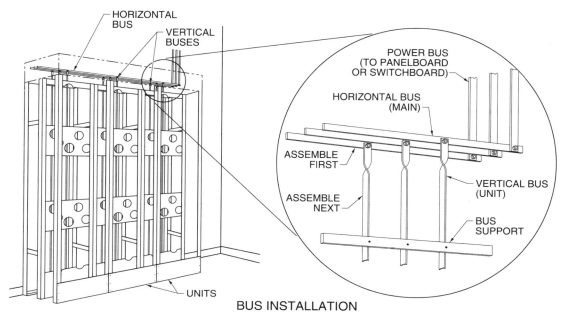

BUS INSTALLATION

Step 4. Pull power and control wire. The two types of wiring in a motor control center are power wiring and control wiring. *Power wiring* is any wire connecting the outputs (motors, heating elements, solenoids, lights, etc.). The size of the power wire depends on the amount of current drawn by each load. *Control wiring* is any wire connecting the inputs (pushbuttons, limit switches, pressure switches, etc.). Control wire is normally No. 14 copper wire.

   Punch the necessary conduit openings and insert conduit fittings before pulling the wire. After the conduit is installed, pull enough wire through to reach the most distant circuit for each run. **See Wire Pulling.** *Note:* Wiring must comply with applicable codes and laws.

3φ FULL-LOAD AMPERES			
HP	**208 V**	**230 V**	**460 V**	**575 V**
½	2.2	2.0	1.0	0.8
1	4.0	3.6	1.8	1.4
2	7.5	6.8	3.4	2.7
5	16.7	15.2	7.6	6.1
10	31.8	28	14	11
20	59.4	54	27	22
30	88	80	40	32
50	143	130	65	52
75	211.2	192	96	77
100	272.8	248	124	99
200	528	480	240	192

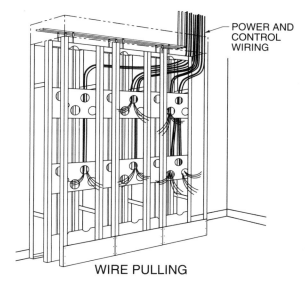

WIRE PULLING

Step 5. Mount starter modules. Individual starter modules are added for each motor. Modules are either fixed or removable. Fixed modules are bolted in. Removable modules are the most common because they are easily inserted and removed. Remove necessary knockouts from starter enclosures. Install wire bushings where desired. Mount the largest starter at the lower left, or the unit farthest from the power wiring first. Wire each module one at a time as required.

To meet NEC® standards, each unit must include a fusible disconnect switch or circuit breaker as the branch circuit protective device. A motor control center normally contains only fusible disconnects or only circuit breakers. **See Starter Module Installation.**

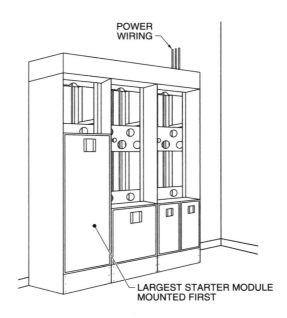

STARTER MODULE INSTALLATION

◻ Application — Busway System Support

> Supports must be placed at the proper distance to prevent excessive deflection when installing busway systems. Manufacturer's specifications provide data for determining proper spacing of busway hangers. **See Load Versus Deflection Specifications.**
>
> For example, a busway weight of 50 lb with hangers spaced 5′ apart causes .70″ of deflection (from Load Versus Deflection Specifications).

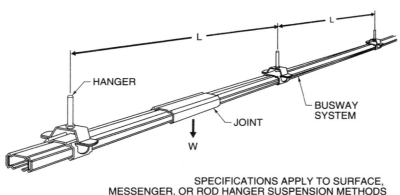

SPECIFICATIONS APPLY TO SURFACE, MESSENGER, OR ROD HANGER SUSPENSION METHODS

LOAD VERSUS DEFLECTION SPECIFICATIONS		
WEIGHT (W) APPLIED AT JOINT (IN LB)	DEFLECTION (IN IN.)	HANGER SPACING (L) (IN FT)
10	.10	
20	.13	
30	.20	$3\frac{1}{3}$
40	.25	
50	.30	
5	.10	
10	.20	
20	.35	5
30	.50	
40	.60	
50	.70	

ACTIVITIES

Name _____ Date _____

○ Activity 10-1. Transformer Tap Connection

Connect the transformer taps for each primary voltage requirement using Primary Tap Connections. Connect the transformer taps for each secondary voltage requirement using Secondary Tap Connections.

PRIMARY TAP CONNECTIONS		
Primary volts	Connect primary lines to	Interconnect
190	H1 & H7	H1 to H6 H2 to H7
200	H1 & H8	H1 to H6 H3 to H8
208	H1 & H9	H1 to H6 H4 to H9
220	H1 &10	H1 to H6 H5 to H10
380	H1 & H7	H2 & H6
400	H1 & H8	H3 & H6
416	H1 & H9	H4 & H6
440	H1 & H10	H5 & H6

SECONDARY TAP CONNECTIONS		
Secondary volts	Connect secondary lines to	Interconnect
220	X1–X4	X2 to X3
110/220	X1–X2–X4	X2 to X3
110	X1–X4	X1 to X3 X2 to X4

1. A transformer has a primary voltage of 416 V. The required secondary voltage is 220 V.

2. A transformer has a primary voltage of 208 V. The required secondary voltage is 110 V.

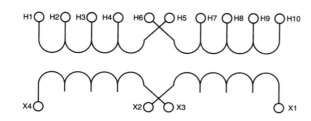

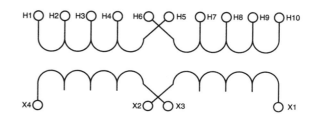

3. A transformer has a primary voltage of 190 V. The required secondary voltage is 110 V.

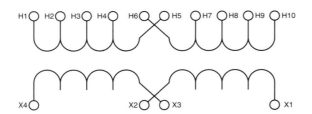

4. A transformer has a primary voltage of 440 V. The required secondary voltage is 220 V.

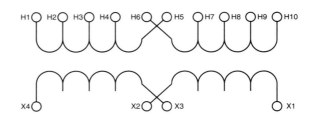

Activity 10-2. Wye and Delta Transformer Connections

Answer questions 1–3 using Transformer Bank 1.

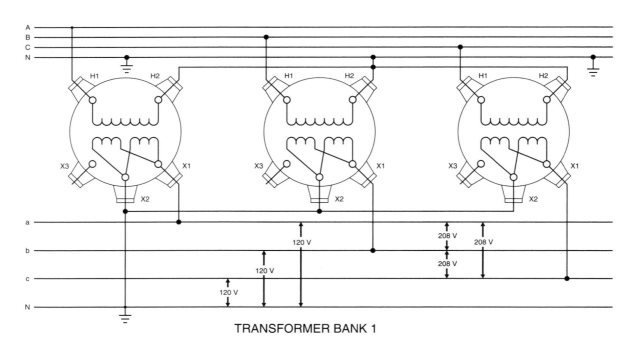

TRANSFORMER BANK 1

_____   **1.** The primary side of the transformer is connected in a(n) _____ configuration.

_____   **2.** The secondary side of the transformer is connected in a(n) _____ configuration.

3. List the power lines to which the loads are connected. If the load cannot be correctly connected to the transformer bank, mark it not possible.

_____   **A.** A 120 VAC motor is connected to _____, _____, or _____.

_____   **B.** A 230 VAC, 1ϕ motor is connected to _____, _____, or _____.

_____   **C.** A dual-voltage (115/230) VAC, 1ϕ motor is connected to _____, _____, or _____ for 115 V.

_____   **D.** A dual-voltage (230/460) VAC, 3ϕ motor is connected to _____, _____, or _____ for 230 V.

_____   **E.** A dual-voltage (230/460) VAC, 3ϕ motor is connected to _____, _____, or _____ for 460 V.

Answer questions 4–6 using Transformer Bank 2.

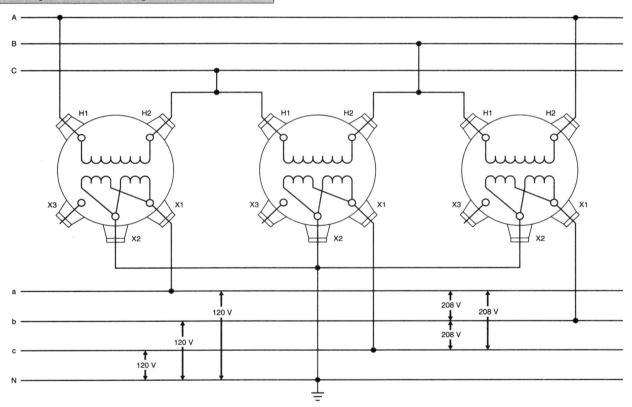

TRANSFORMER BANK 2

_____   **4.** The primary side of the transformer is connected in a(n) _____ configuration.

_____   **5.** The secondary side of the transformer is connected in a(n) _____ configuration.

6. List the power lines to which the loads are connected. If the load cannot be correctly connected to the transformer bank, mark it not possible.

_____   **A.** A 120 VAC motor is connected to _____, _____, or _____.

_____   **B.** A 230 VAC, 1ϕ motor is connected to _____, _____, or _____.

_____ **C.** A dual-voltage (115/230) VAC motor is connected to _____, _____, or _____ for 115 V.

_____ **D.** A dual-voltage (115/230) VAC motor is connected to _____, _____, or _____ for 230 V.

_____ **E.** A dual-voltage (230/460) VAC, 3ϕ motor is connected to _____ for 460 V.

Answer questions 7–9 using Transformer Bank 3.

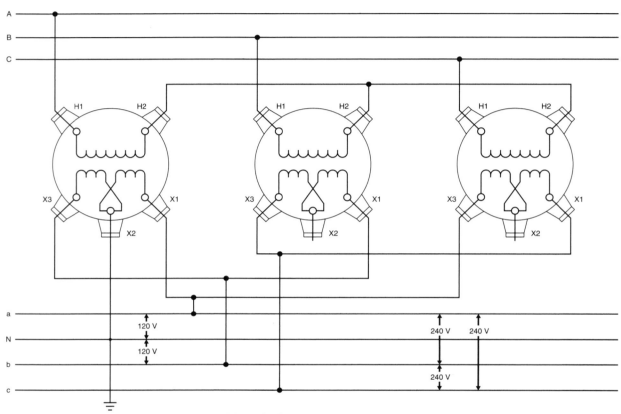

TRANSFORMER BANK 3

_____ **7.** The primary side of the transformer is connected in a(n) _____ configuration.

_____ **8.** The secondary side of the transformer is connected in a(n) _____ configuration.

9. List the power lines to which the loads are connected. If the load cannot be correctly connected to the transformer bank, mark it not possible.

_____ **A.** A 120 VAC motor is connected to _____ or _____.

_____ **B.** A 230 VAC, 1ϕ motor is connected to _____, _____, or _____.

_____ **C.** A dual-voltage (115/230) VAC motor is connected to _____, _____, or _____ for 115 V.

_____ **D.** A dual-voltage (115/230) VAC, 3ϕ motor is connected to _____, _____, or _____ for 230 V.

_____ **E.** A dual-voltage (230/460) VAC, 3ϕ motor is connected to _____, _____, or _____ for 460 V.

Answer questions 10–12 using Transformer Bank 4.

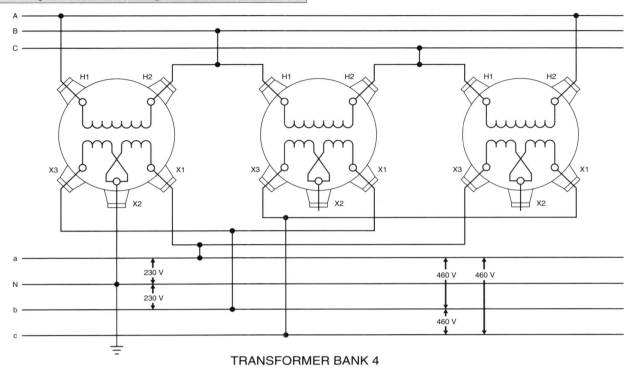

TRANSFORMER BANK 4

_____ **10.** The primary side of the transformer is connected in a _____ configuration.

_____ **11.** The secondary side of the transformer is connected in a _____ configuration.

12. List the power lines to which the loads are connected. If the load can not be correctly connected to the transformer bank, mark it not possible.

_____ **A.** A 120 VAC motor is connected to _____.

_____ **B.** A 230 VAC, 1φ motor is connected to _____ or _____.

_____ **C.** A dual-voltage (115/230) VAC motor is connected to _____, _____, or _____ for 115 V.

_____ **D.** A dual-voltage (115/230) VAC motor is connected to _____, _____, or _____ for 230 V.

_____ **E.** A dual-voltage (230/460) VAC, 3φ motor is connected to _____, _____, or _____ for 230 V.

_____ **F.** A dual-voltage (230/460) VAC, 3φ motor is connected to _____, _____, or _____ for 460 V.

◯ Activity 10-3.   Motor Control Center Assembly

Answer the questions using Motor Control Center Assembly Procedure on pages 88–90.

1. What considerations should be given when selecting a location for the control center?

2. How should the motor control center be secured to prevent movement?

3. How are the busbars of adjacent buses connected?

4. How much wire is required in the motor control center for the preliminary wiring?

5. In what order should the starters be mounted in the control center?

○ Activity 10-4. Busway System Support

Answer the questions using Load Versus Deflection Specifications on page 90.

_____ 1. A busway weight of 30 lb with hangers spaced 5′ apart causes _____″ of deflection.

_____ 2. A busway weight of 30 lb with hangers spaced 3.33′ apart causes _____″ of deflection.

3. What is the hanger spacing if a maximum deflection of no more than .18″ is allowed?

    A. 5 lb

_____ B. 10 lb

_____ C. 15 lb

_____ D. 20 lb

_____ E. 30 lb

_____ F. 40 lb

_____ G. 50 lb

4. What is the hanger spacing if a maximum deflection of no more than .65″ is allowed?

_____ A. 5 lb

_____ B. 10 lb

_____ C. 15 lb

_____ D. 20 lb

_____ E. 30 lb

_____ F. 40 lb

_____ G. 50 lb

_____ 5. A maximum of _____ lb/ft of a busway can be connected to the hangers if the busway system is supported every 5′ and no more than .70″ of deflection is allowed.

Application — Digital Circuit Logic

Digital logic circuits make decisions in control circuits. A *digital signal* is a signal represented by one of two states. The signal is either high (1) or low (0). A high signal is normally 5 V, but can be from 2.4 V to 5 V. A low signal is normally 0 V, but can be from 0 V to .8 V. Digital logic gates are used to control electrical circuits. The AND, OR, and NOT logic gates are the three basic logic functions that make up most digital circuit logic. The NOT gate is used to invert the incoming signal to the gate. The NOR gate is a NOT, OR, or inverted OR gate. The NAND gate is a NOT, AND, or inverted AND gate. AND, OR, NOT, NOR, and NAND logic has the same meaning for digital logic, hard-wired electrical logic, and relay logic. **See Basic Logic Functions.**

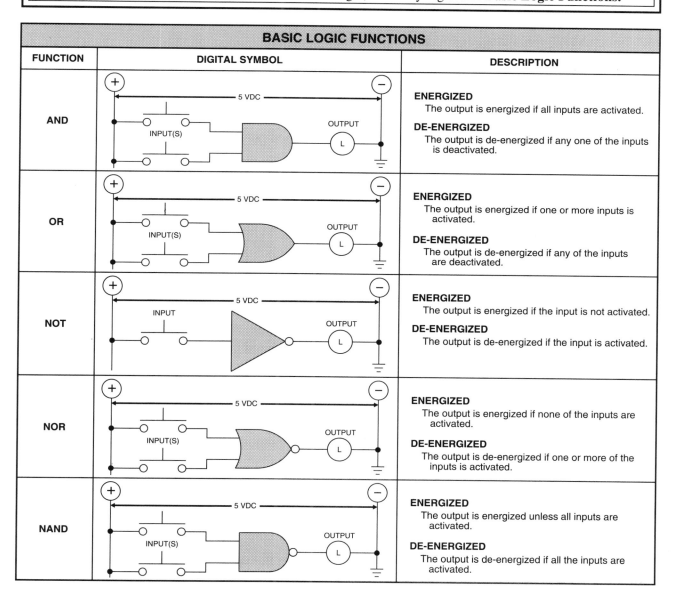

FUNCTION	DIGITAL SYMBOL	DESCRIPTION
	BASIC LOGIC FUNCTIONS	
AND		**ENERGIZED** The output is energized if all inputs are activated. **DE-ENERGIZED** The output is de-energized if any one of the inputs is deactivated.
OR		**ENERGIZED** The output is energized if one or more inputs is activated. **DE-ENERGIZED** The output is de-energized if any of the inputs are deactivated.
NOT		**ENERGIZED** The output is energized if the input is not activated. **DE-ENERGIZED** The output is de-energized if the input is activated.
NOR		**ENERGIZED** The output is energized if none of the inputs are activated. **DE-ENERGIZED** The output is de-energized if one or more of the inputs is activated.
NAND		**ENERGIZED** The output is energized unless all inputs are activated. **DE-ENERGIZED** The output is de-energized if all the inputs are activated.

☐ Application — Solid-state Relays

Because the output of a digital system is either 0 V or 5 VDC, an interface is required when operating a higher or different type of voltage. For example, a solid-state relay is used as an interface when digital circuits control higher DC outputs, such as 12 V, 24 V, or 36 VDC solenoids and higher AC outputs such as 120 V or 240 VAC motor starters. However, the output of a solid-state relay is only used for AC or DC. **See Solid-state Relays.**

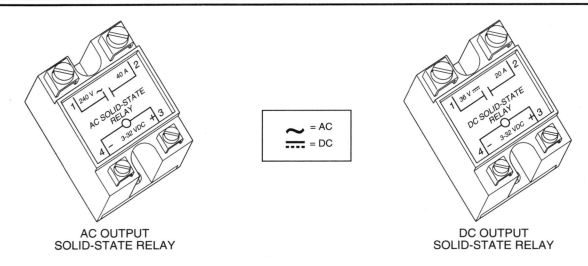

AC OUTPUT
SOLID-STATE RELAY

DC OUTPUT
SOLID-STATE RELAY

☐ Application — Digital AND Circuit Logic

An *AND gate* is a digital logic device that has two or more inputs and one output. The logic decision of an AND gate is based on the status of the inputs. If both or all inputs are high, the output is high. AND circuit logic is the same for hard-wired electrical circuits using two normally open contacts and digital circuits using AND gates.

The digital logic control circuit performs the same function as the electrical control circuit. The electrical control circuit uses 120 VAC and the digital control circuit uses 5 VDC for the control voltage. **See AND Logic.**

Note: The pushbuttons to all logic gates are normally connected to produce a low-voltage signal when the pushbuttons are open. The pushbuttons are drawn to simplify their understanding.

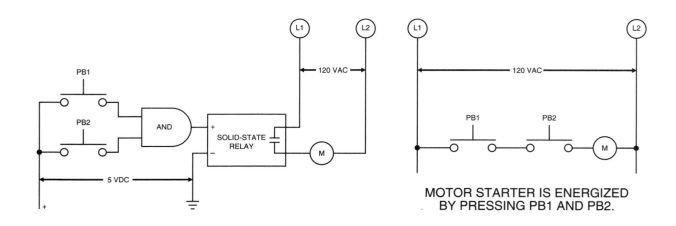

MOTOR STARTER IS ENERGIZED
BY PRESSING PB1 AND PB2.

Application — Digital OR Circuit Logic

An *OR gate* is a digital logic device that has two or more inputs and one output. The logic decision of an OR gate is based on the status of the inputs. If any of the inputs are high, the output is high. OR circuit logic is the same for both hard-wired electrical circuits using two normally open contacts and digital circuits using OR gates. **See OR Logic.**

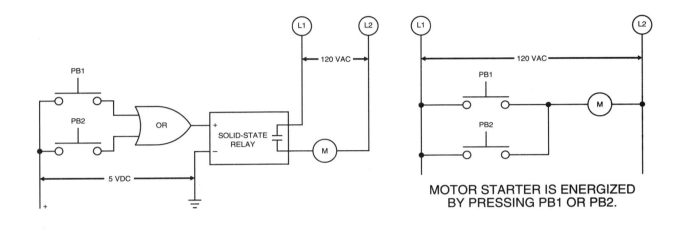

MOTOR STARTER IS ENERGIZED
BY PRESSING PB1 OR PB2.

Application — Digital NOT Circuit Logic

A *NOT gate* is a digital logic device that has one input and one output. The output of a NOT gate is the opposite of the input. If the input is low, the output is high. If the input is high, the output is low. NOT circuit logic is the same for both hard-wired electrical circuits using normally closed contacts and digital circuits using a NOT gate. **See NOT Logic.** A circle placed at the front of the logic gates inverts the output signal.

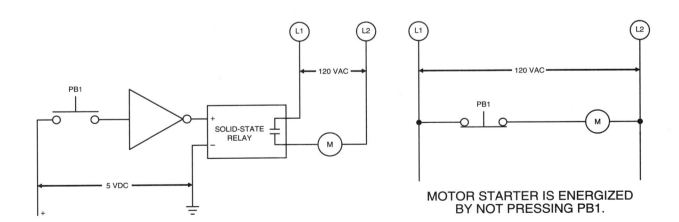

MOTOR STARTER IS ENERGIZED
BY NOT PRESSING PB1.

◻ Application — Digital NOR Circuit Logic

A *NOR gate* is a digital logic device that has two or more inputs and one output. The logic decision of a NOR gate is based on the status of the inputs. If any one or all inputs are high (ON), the output is low (OFF). NOR circuit logic is the same for hard-wired electrical circuits using two normally closed contacts connected in series and digital circuits using NOR gates. **See NOR Logic.**

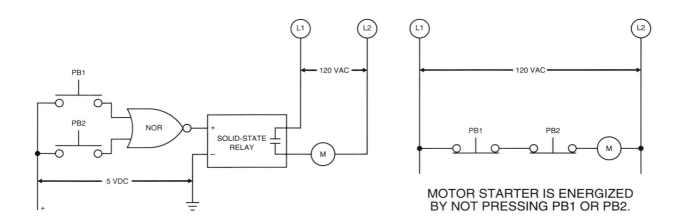

MOTOR STARTER IS ENERGIZED
BY NOT PRESSING PB1 OR PB2.

◻ Application — Digital NAND Circuit Logic

A *NAND gate* is a digital logic device that has two or more inputs and one output. The logic decision of a NAND gate is based on the status of the inputs. If any one or all inputs are low (OFF), the output is high (ON). NAND circuit logic is the same for hard-wired electrical circuits using two normally closed contacts connected in parallel and digital circuits using NAND gates. **See NAND Logic.**

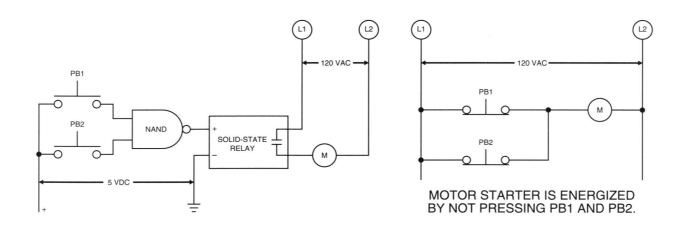

MOTOR STARTER IS ENERGIZED
BY NOT PRESSING PB1 AND PB2.

Application — Combination Logic

Digital logic gates are combined to develop different logic functions. By combining the basic logic gates, digital circuits can solve almost any control application. This includes all basic hard-wired industrial circuits such as start/stop, jogging, forward/reversing, etc. Digital circuits can also be used to solve switching applications that are not practical using hard-wired circuits.

Start/Stop Control Circuit

Start/stop circuits require memory. In a memory circuit, the load (usually a motor starter) remains energized after a start pushbutton is pressed and released. The load remains energized until a stop pushbutton is pressed. The circuit has memory because the load remains energized after the start pushbutton is released. **See Start/stop Control Circuit.**

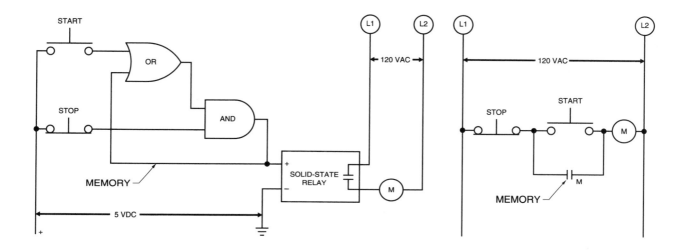

Jogging Control Circuit

Jogging is a control function that allows for frequent starting and stopping of a motor for short periods of time. In a jogging circuit, the motor is energized only when the jog pushbutton is held down. Jogging is used for positioning the driven load by moving the driven load a small distance each time the motor starts. **See Jogging Control Circuit.**

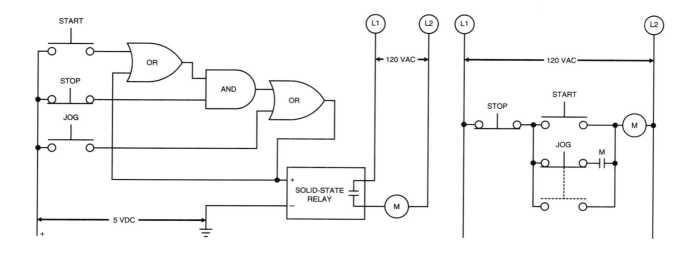

Coin Changer

AND and OR gates can be combined to produce the logic required in a basic coin changer. One set of inputs is activated as coins are inserted into the vending machine. Another set of inputs are activated when the cost of the product is determined. If the cost of the product selected is less than the coin inserted, the digital circuit makes the decision required to give the correct change.

Note: It is assumed that only one coin is inserted for the purchase of the product, and the cost of the product selected is either 5, 10, 15, 20, or 25 cents.

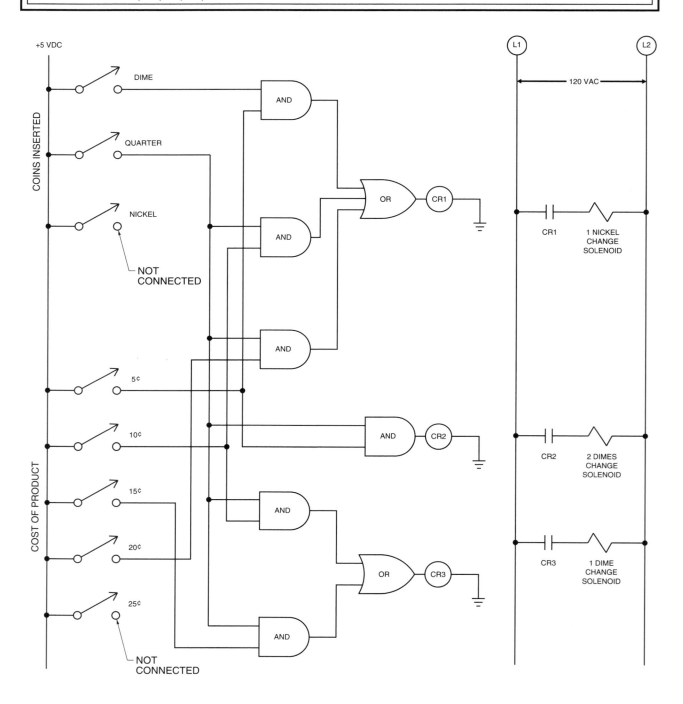

ACTIVITIES

Name _____   Date _____

⬤ Activity 11-1.   Digital AND Circuit Logic

Wire the pushbuttons to control the motor using the solid-state relay as the interface. Connect the circuit so the motor starts only when Pushbutton 1 and Pushbutton 2 are pressed. Connect the pushbuttons and solid-state relay to the correct terminal blocks. Make all connections at the terminal blocks. Each terminal block screw is designed to hold up to three wires.

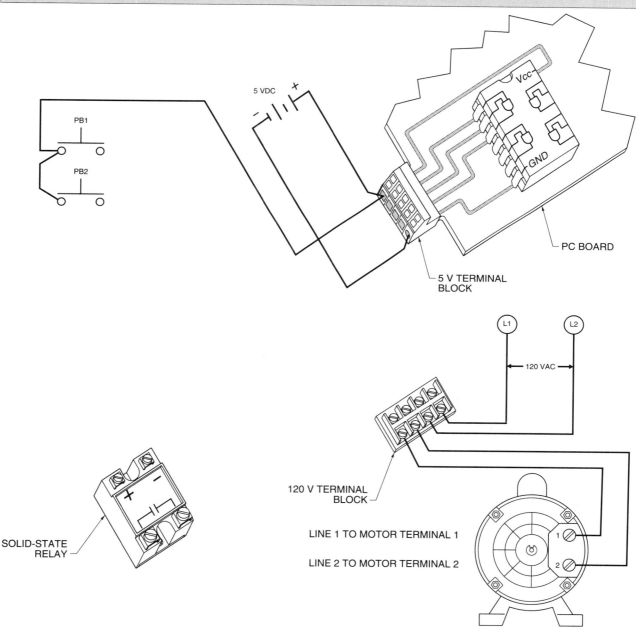

PB1

PB2

5 VDC

Vcc

GND

PC BOARD

5 V TERMINAL BLOCK

L1   L2

120 VAC

120 V TERMINAL BLOCK

SOLID-STATE RELAY

LINE 1 TO MOTOR TERMINAL 1

LINE 2 TO MOTOR TERMINAL 2

1

2

Activity 11-2. Digital MEMORY Circuit Logic

Wire the pushbuttons to control the motor using the solid-state relay as the interface. Connect the circuit so the motor starts only when Pushbutton 1 and Pushbutton 2 are pressed or when Pushbutton 3 is pressed. Connect the pushbuttons and solid-state relay to the correct terminal blocks. Make all connections at the terminal blocks. Each terminal block screw is designed to hold up to three wires.

Note: Connect only one output from a gate to the relay. Do not connect gate outputs in parallel.

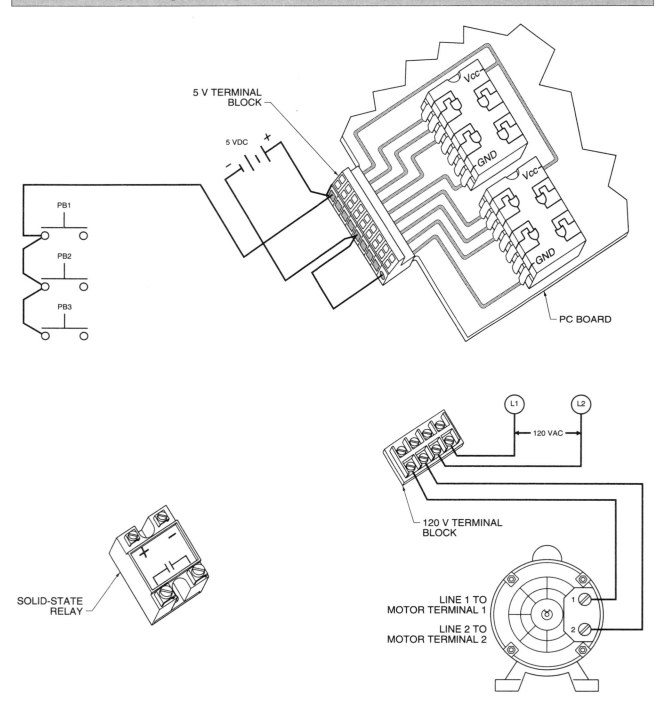

⬤ Activity 11-3.   Digital NOT Circuit Logic

Wire the pushbuttons to control the motor using the solid-state relay as the interface. Connect the circuit so the motor starts only when Pushbutton 1 is pressed and the limit switch is not activated. Connect the pushbuttons and solid-state relay to the correct terminal blocks. Make all connections at the terminal blocks. Each terminal block screw is designed to hold up to three wires.

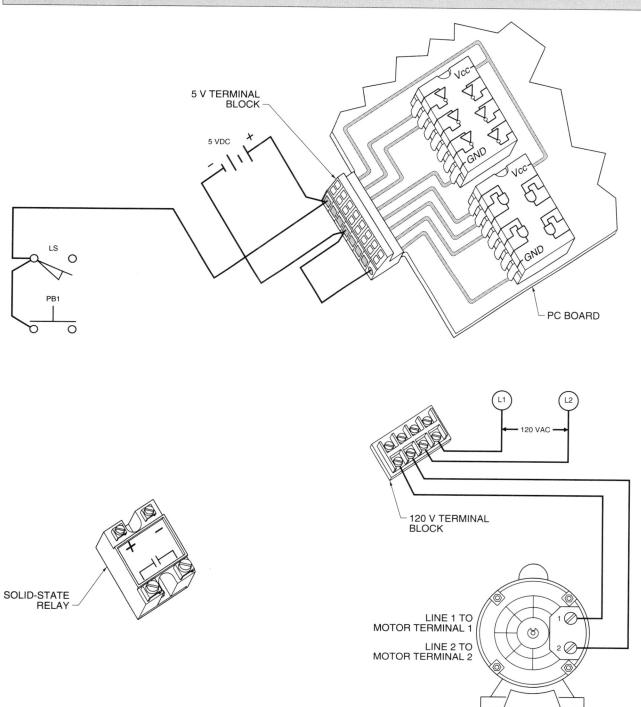

Activity 11-4. Digital Combination Circuit Logic

Wire the pushbuttons to control the motor using the solid-state relay as the interface. Connect the circuit so the motor starts only when the start pushbutton is pressed and released and stops when the stop pushbutton is pressed. Connect the pushbuttons and solid-state relay to the correct terminal blocks. Make all connections at the terminal blocks. Each terminal block screw is designed to hold up to three wires.

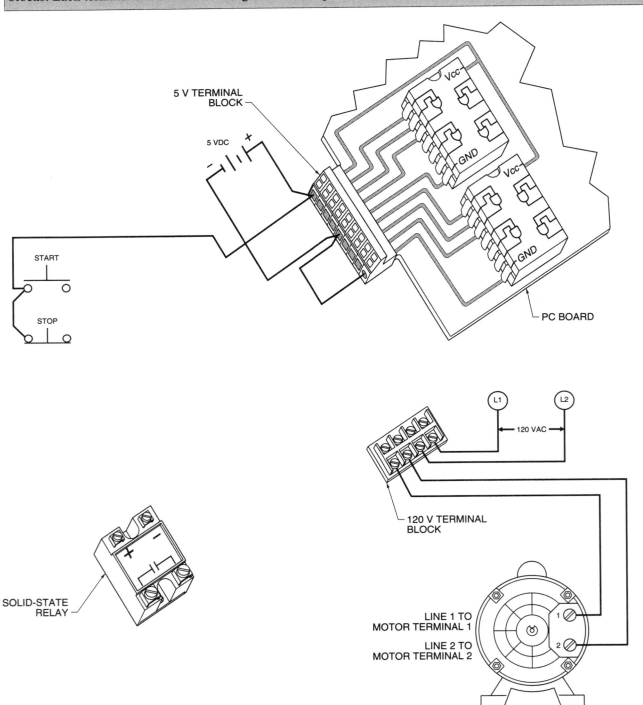

APPLICATIONS

▢ Application — Contact Arrangement Identification

The different contact arrangements of relays and timers are shown in schematic diagrams which describe the poles, throws, and breaks of the relay or timer. Electricians must identify contacts when installing, replacing, or specifying relays. **See Relay Contacts.**

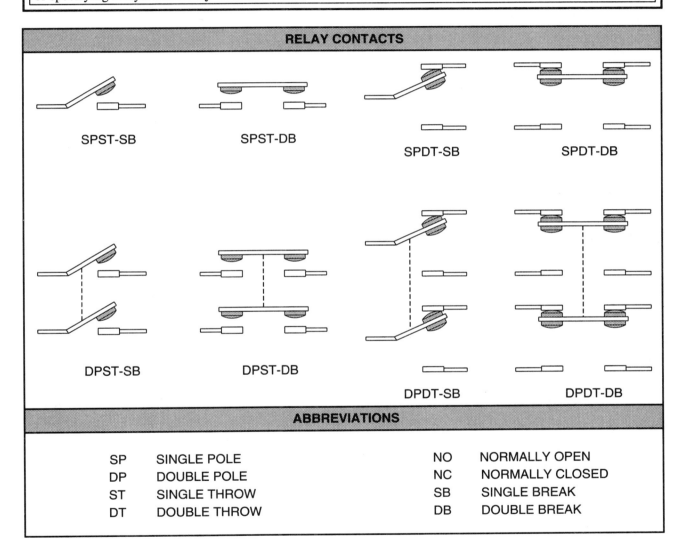

RELAY CONTACTS

SPST-SB SPST-DB SPDT-SB SPDT-DB

DPST-SB DPST-DB DPDT-SB DPDT-DB

ABBREVIATIONS

SP	SINGLE POLE		NO	NORMALLY OPEN
DP	DOUBLE POLE		NC	NORMALLY CLOSED
ST	SINGLE THROW		SB	SINGLE BREAK
DT	DOUBLE THROW		DB	DOUBLE BREAK

▢ Application — Ordering General-purpose Relays

Manufacturers use numbers to identify the different types of relays. These numbers include the dimensions, electrical features, mechanical features, and contact arrangements of the relay. **See Relay Specifications.**

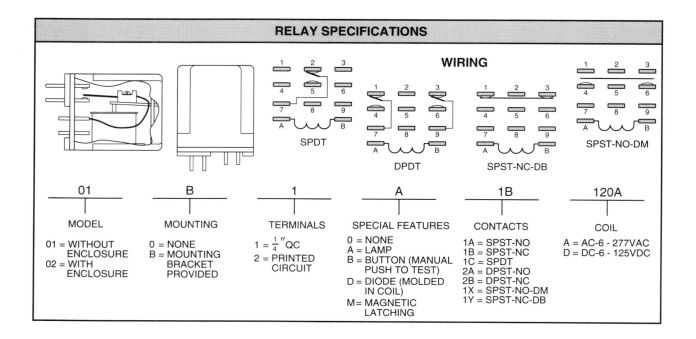

RELAY SPECIFICATIONS

WIRING

SPDT

DPDT

SPST-NC-DB

SPST-NO-DM

01	B	1	A	1B	120A
MODEL	MOUNTING	TERMINALS	SPECIAL FEATURES	CONTACTS	COIL

01 = WITHOUT ENCLOSURE
02 = WITH ENCLOSURE

0 = NONE
B = MOUNTING BRACKET PROVIDED

$1 = \frac{1}{4}''$ QC
2 = PRINTED CIRCUIT

0 = NONE
A = LAMP
B = BUTTON (MANUAL PUSH TO TEST)
D = DIODE (MOLDED IN COIL)
M = MAGNETIC LATCHING

1A = SPST-NO
1B = SPST-NC
1C = SPDT
2A = DPST-NO
2B = DPST-NC
1X = SPST-NO-DM
1Y = SPST-NC-DB

A = AC-6 - 277VAC
D = DC-6 - 125VDC

Application — Magnetic Latching Relays

Control circuit memory is developed with holding contacts connected in parallel with the control circuit start button or by using a latching relay.

When using holding contacts connected in parallel with the start button, the load is energized when the start button is pressed and released. The load remains energized until the memory circuit is broken. The memory circuit is normally broken by a second stop button. The memory circuit is also broken when a voltage failure occurs. The circuit does not turn ON when the voltage is returned. This circuit is a low-voltage (undervoltage) protection circuit because the memory circuit is broken on low voltage.

When using a latching relay to develop circuit memory, the latching relay provides a holding function through a set/reset coil on the relay. When voltage is applied to the set coil, the relay contacts change position. The contacts stay in this position even if the voltage is removed. To change the contacts back, voltage must be applied to the reset coil. **See Latching Relay Wiring Diagrams.**

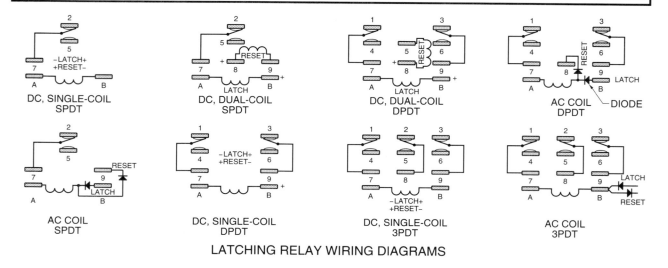

DC, SINGLE-COIL SPDT

DC, DUAL-COIL SPDT

DC, DUAL-COIL DPDT

AC COIL DPDT

AC COIL SPDT

DC, SINGLE-COIL DPDT

DC, SINGLE-COIL 3PDT

AC COIL 3PDT

LATCHING RELAY WIRING DIAGRAMS

Application — Heat Sink Selection

Solid-state relays use heat sinks to prevent damage to the switching element of the relay. The size of a heat sink is determined by dividing the temperature rise at the relay by the power at the relay contact. This ratio is the required thermal resistance of the heat sink (in °C/W) and is used when rating and selecting a heat sink. To select a heat sink for a relay, apply the procedure:

Step 1. Determine the current of the relay controls. If the current is not known, apply Ohm's Law.

Step 2. Determine the ambient temperature of the relay installation.

Step 3. Determine the temperature rise at the relay. To calculate temperature rise, apply the formula:

$$T_R = M_{TR} - A_T \quad \text{where}$$

T_R = temperature rise at relay (in °C) A_T = ambient temperature at relay (in °C)

M_{TR} = maximum temperature rise at relay (normally 110°C)

Step 4. Determine the power drop at relay contacts. To calculate power drop, apply the formula:

$$P = I_L \times V_D \quad \text{where}$$

P = power drop at relay contact (in watts) I_L = load current (in amps)

V_D = voltage drop at relay contact (normally 1.6 V)

Step 5. Determine the thermal resistance of the heat sink. The thermal resistance of a heat sink in calculated or found on relay specifications. To calculate the thermal resistance, apply the formula:

$$R_{TH} = \frac{T_R}{P} \quad \text{where}$$

R_{TH} = thermal resistance of heat sink (in °C/W) T_R = temperature rise at relay (in °C)

P = power drop at relay contact (in watts)

Step 6. Select the heat sink type with a value equal to or smaller than the required size. **See Heat Sink Selections.**

Example: Heat Sink Selection — Calculation

A relay controls a 25 A load in a 50°C ambient installation. Find the required heat sink.

Step 1. Determine the current the relay controls.
The relay controls a 25 A load.

Step 2. Determine the ambient temperature of the relay installation.
The relay is installed in a 50°C ambient temperature location.

Step 3. Determine the temperature rise at the relay.

$T_R = M_{TR} - A_T$ $T_R = 110 - 50$
$T_R = \textbf{60°C}$

Step 4. Determine the power drop at relay contacts.

$P = I_L \times V_D$ $P = 25 \times 1.6$
$P = \textbf{40 W}$

Step 5. Determine the thermal resistance of the heat sink.

$R_{TH} = \dfrac{T_R}{P}$ $R_{TH} = \dfrac{60}{40}$

$R_{TH} = \textbf{1.5°C/W}$

HEAT SINK SELECTIONS		
TYPE	H x W x L (mm)	R_{TH} (°C/W)
01	15 x 79 x 100	2.5
02	15 x 100 x 100	2.0
03	25 x 97 x 100	1.5
04	37 x 120 x 100	0.9
05	40 x 160 x 150	0.5
06	40 x 200 x 150	0.4

Step 6. Select a heat sink with a thermal resistance value equal to or smaller than the required size (from Heat Sink Selections). Minimum selection for 1.5°C/W is **Type 03.** Best selection for 0.9°C/W is **Type 04.**

Heat Sink Selection—Relay Specifications

To find the thermal resistance of a heat sink using manufacturer's relay specifications, find the intersection of the load current and ambient temperature on a relay specification. The thermal resistance value is represented by the °C/W line directly above the intersection of the load current and ambient temperature. **See 40 A Relay Specifications.**

Example: Heat Sink Selection — Relay Specifications

A 40 A relay controls a 35 A load in a 40°C ambient temperature installation. Find the required heat sink.

Step 1. Determine the current that the relay controls.

The relay controls a 35 A load.

Step 2. Determine the ambient temperature of the relay installation.

The relay is installed in a 40°C ambient temperature location.

Step 3. Determine thermal resistance of the heat sink.

Find the intersection of the 35 A and 40°C lines on the relay specifications. The thermal resistance line directly above the intersection point is 0.9°C/W.

Step 4. Select the heat sink type with a value equal to or smaller than the required size (from Heat Sink Selections).

Minimum selection for 0.9°C/W is **Type 04.**
Best selection for 0.5°C/W is **Type 05.**

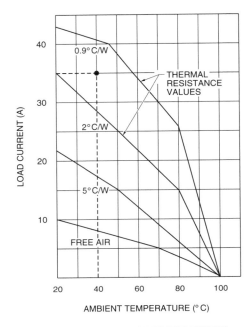

40 A RELAY SPECIFICATIONS

▢ Application — Solid-state Relay Installation

When installing solid-state relays, allow maximum heat transfer from the relay. Maximum heat transfer depends on how well the heat produced inside the relay is transferred to the heat sink, how well the heat transferred to the heat sink is transferred to the surrounding air, and how well the surrounding air is replaced with cooler air.

Thermal conductive compound is used to ensure proper heat transfer from the relay to the heat sink. *Thermal conductive compound* (thermal paste) is used to reduce the high thermal resistance of the air gap between the heat sink and the solid-state relay. Thermal conductive compound provides an excellent thermal path for transferring heat from the solid-state relay to the heat sink. Thermal conductive compound is spread evenly over the entire surfaces to be joined. When replacing relays, clean the old compound before applying new compound.

Heat Sink Mounting

A heat sink must be properly mounted and placed in a system to ensure proper heat transfer from the heat sink to the surrounding air. Proper mounting includes mounting the fins of the heat sink vertically to ensure the heated air is moved away form the relay. Proper placing includes mounting the heat sinks to permit free air flow from natural convection or forced air flow.

ACTIVITIES

Name _____   Date _____

◯ Activity 12-1.   Contact Arrangement Identification

Identify the number of poles, throws, and breaks for each contact arrangement. State the coil pins, normally open contacts, and normally closed contacts.

_____   **1.** Number of poles
_____   **2.** Number of throws
_____   **3.** Number of breaks
_____   **4.** Coil pins
_____   **5.** Normally open contacts
_____   **6.** Normally closed contacts

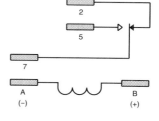

_____   **7.** Number of poles
_____   **8.** Number of throws
_____   **9.** Number of breaks
_____   **10.** Coil pins
_____   **11.** Normally open contacts
_____   **12.** Normally closed contacts

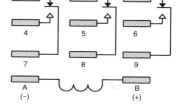

_____   **13.** Number of poles
_____   **14.** Number of throws
_____   **15.** Number of breaks
_____   **16.** Coil pins
_____   **17.** Normally open contacts
_____   **18.** Normally closed contacts

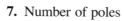
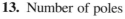

_____   **19.** Number of poles
_____   **20.** Number of throws
_____   **21.** Number of breaks
_____   **22.** Coil pins
_____   **23.** Normally open contacts
_____   **24.** Normally closed contacts

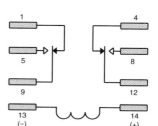

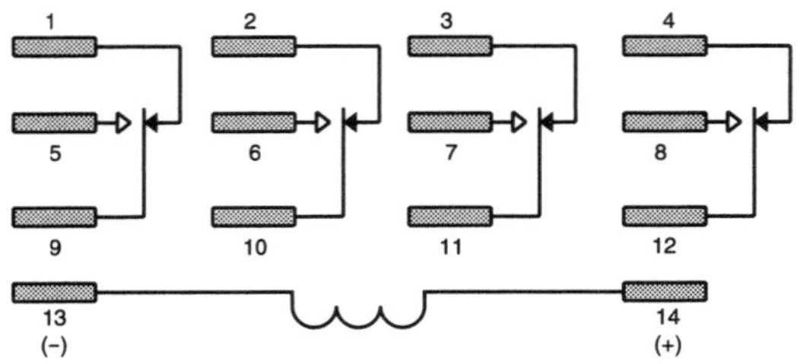

_____ **25.** Number of poles
_____ **26.** Number of throws
_____ **27.** Number of breaks
_____ **28.** Coil pins
_____ **29.** Normally open contacts
_____ **30.** Normally closed contacts

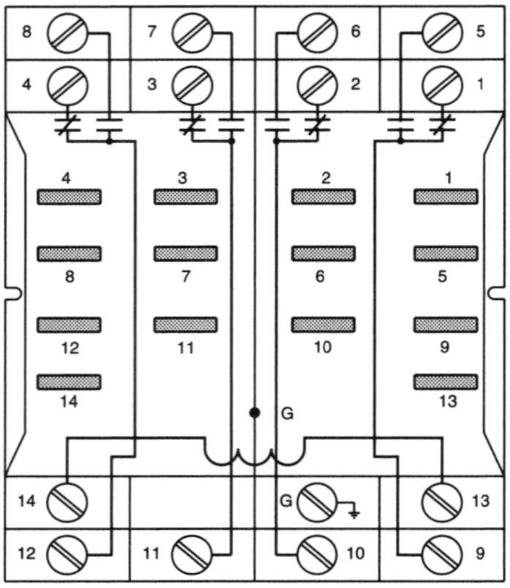

_____ **31.** Number of poles
_____ **32.** Number of throws
_____ **33.** Number of breaks
_____ **34.** Coil pins
_____ **35.** Normally open contacts
_____ **36.** Normally closed contacts

⭘ Activity 12-2. Ordering General-purpose Relays

Answer the questions using Relay Specifications on page 108.

1. A relay has a model number 01-B-1-B-1C-120A.

_____ **A.** The relay has _____ normally open contacts.
_____ **B.** The terminal numbers of the contacts are _____ .
_____ **C.** The relay has _____ normally closed contacts.
_____ **D.** The terminal numbers of the contacts are _____ .
_____ **E.** The terminal numbers of the coil are _____ .
_____ **F.** The rating of the coil is _____ VAC.

2. Identify the relay types.

A. SPDT

B. DPDT

C. SPST-NC-DB

D. SPST-NC-DM

_____ 3. The SPST-NC-DB relay has _____ normally closed contacts.
_____ 4. The SPST-NO-DB relay has _____ normally open contacts.
5. A relay has a model number 01-B-1-A-2A-120A. When is the lamp energized?

◯ Activity 12-3.   Magnetic Latching Relays

1. Draw the line diagram of the Motor Jog Circuit including a relay that allows the motor to be jogged or started.

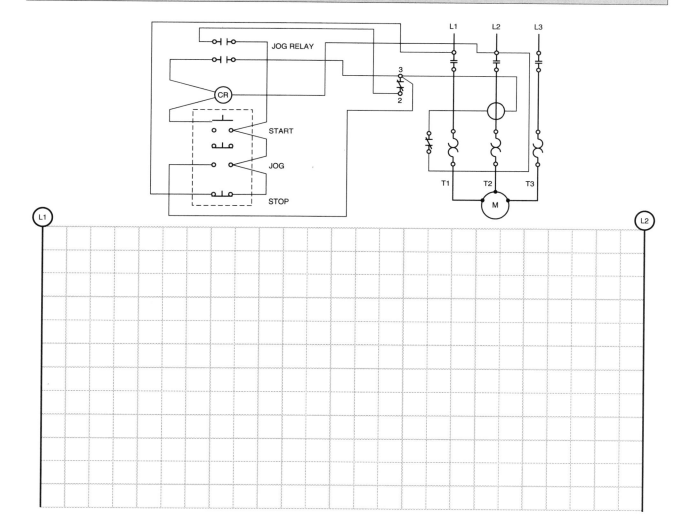

2. In the Motor Jog Circuit on page 113 using the jog relay, the starter is de-energized when a power failure occurs. The starter does not automatically energize when power is reapplied. To automatically energize a starter after a power failure, a latching relay is used.

Redraw the line diagram of the circuit using a jog relay and add an AC coil DPDT latching relay in place of the jog relay. Change the stop button to a normally open stop button. When the start button is used, the starter energizes and remains energized after the start button is pressed and released. When the jog button is used, the starter energizes only as long as the jog button is pressed.

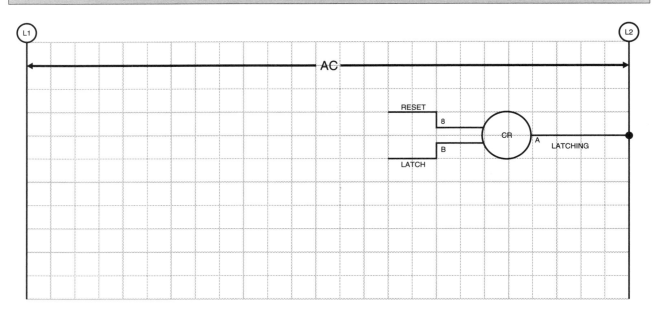

3. Draw a line diagram using the DC, dual-coil, SPST latching relay. An ON button energizes a lamp and a second OFF button de-energizes the lamp. Mark each terminal number used from the latching relay on the line diagram.

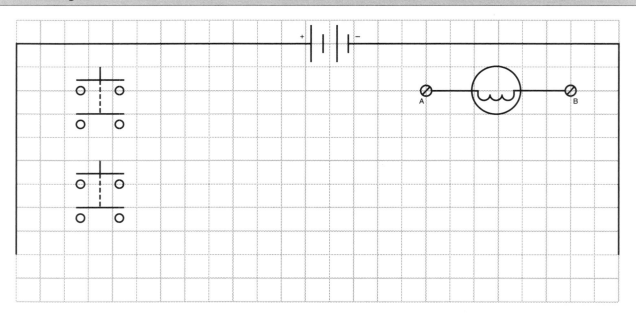

4. Draw a line diagram using the AC coil, 3PDT latching relay. An ON button energizes a motor starter in the AC circuit and a second OFF button de-energizes the starter. Use the first set of contacts (1, 4, 7) to control the starter. Use the second set of contacts to energize a solenoid in the DC circuit every time the starter is energized. Use the third set of contacts to de-energize a light in the DC circuit every time the starter is energized. Mark each terminal number used from the latching relay on the line diagram.

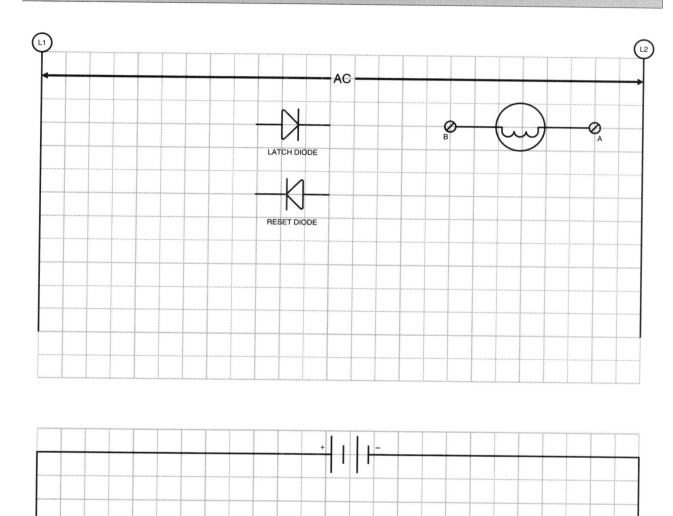

Activity 12-4.   Heat Sink Selection

Select the correct size heat sink for the applications using Heat Sink Selections on page 109.

_____
1. A 120 V, 30 A load is controlled by a relay installed in an ambient temperature of 40°C. A type _____ heat sink is required.

_____
2. A 120 V, 8 Ω load is controlled by a relay installed in an ambient temperature of 75°C. A type _____ heat sink is required.

_____
3. A 240 V, 2500 W load is controlled by a relay installed in an ambient temperature of 55°C. A type _____ is required.

_____
4. A 240 V, 10 Ω load is controlled by a relay installed in an ambient temperature of 45°C. A type _____ heat sink is required.

_____
5. A 240 V, 9.6 Ω, 6000 W load is controlled by a relay installed in an ambient temperature of 60°C. A type _____ heat sink is required.

Activity 12-5.   Heat Sink Installation

Mark each relay/heat sink mounting method as correct or incorrect.

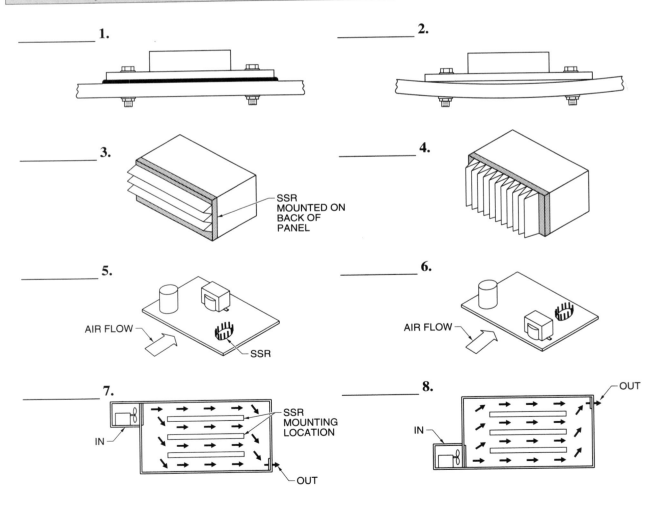

Application — Proximity Sensors

Proximity Sensors

Proximity sensors are electrical switches that do not require physical contact for activation. The three types of proximity sensors are inductive, capacitive, and photoelectric.

Inductive sensors detect metallic objects. Nominal sensing distances range from .5 mm to 50 mm. The maximum sensing distance depends on the size of the object to be sensed and the type of metal. For example, iron is sensed at twice the distance of aluminum. Applications for inductive sensors include positioning, fan blade detection, drill bit breakage, and solid-state replacement of mechanical limit switches. **See Inductive Sensor.**

INDUCTIVE SENSOR

SPECIFICATIONS

SUPPLY VOLTAGE	10 VDC- 40 VDC
AMBIENT TEMPERATURE	–20° C TO +60° C
TARGET	FERROUS METALS
SENSING DISTANCE	40 mm
LEAKAGE CURRENT	1.7 mA
RESPONSE TIME	150 Hz
OUTPUT	MINIMUM: 20 mA MAXIMUM: 200 mA

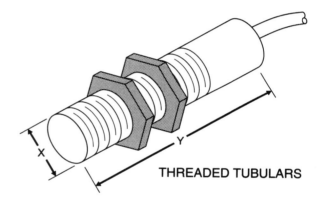

THREADED TUBULARS

SIZE (mm)	TYPE						
	A	B	C	D	E	F	G
X	6	8	12	16	18	20	30
Y	35	72	42	80	42	80	50

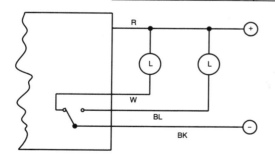

**NPN TRANSISTOR OUTPUT
(CURRENT SINK)**

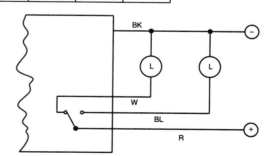

**PNP TRANSISTOR OUTPUT
(CURRENT SOURCE)**

Capacitive Sensors

Capacitive sensors detect conductive and nonconductive solid, fluid, or granulated substances. Nominal sensing distances range from 3 mm to 20 mm. The maximum sensing distance depends on the physical and electrical characteristics (dielectric) of the object to be detected. Materials with larger dielectric numbers are easier to detect with a capacitive sensor. Applications for capacitive sensors include sensing the level of products, such as sugar, grain, and sand in a container. **See Capacitive Sensor.**

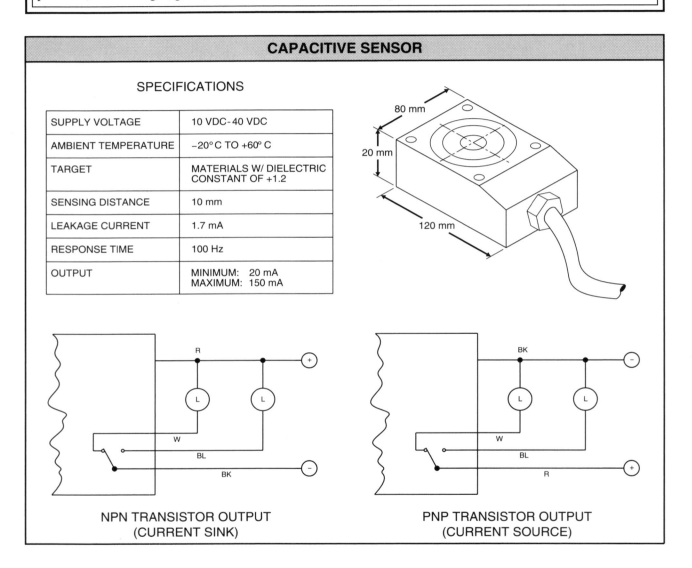

CAPACITIVE SENSOR

SPECIFICATIONS

SUPPLY VOLTAGE	10 VDC- 40 VDC
AMBIENT TEMPERATURE	–20° C TO +60° C
TARGET	MATERIALS W/ DIELECTRIC CONSTANT OF +1.2
SENSING DISTANCE	10 mm
LEAKAGE CURRENT	1.7 mA
RESPONSE TIME	100 Hz
OUTPUT	MINIMUM: 20 mA MAXIMUM: 150 mA

NPN TRANSISTOR OUTPUT
(CURRENT SINK)

PNP TRANSISTOR OUTPUT
(CURRENT SOURCE)

Photoelectric Sensors

Photoelectric sensors detect most materials and have the longest sensing distance. Nominal sensing distances range to over 150 cm for sensors that do not use a separate reflector or receiver. Photoelectric sensors that include a separate receiver are available in ranges over 100′. The maximum sensing distance depends on the color and type of surface of the reflecting object. Applications using photoelectric sensors include detecting almost any object moving along a conveyor system and no-touch detection. **See Photoelectric Sensor.**

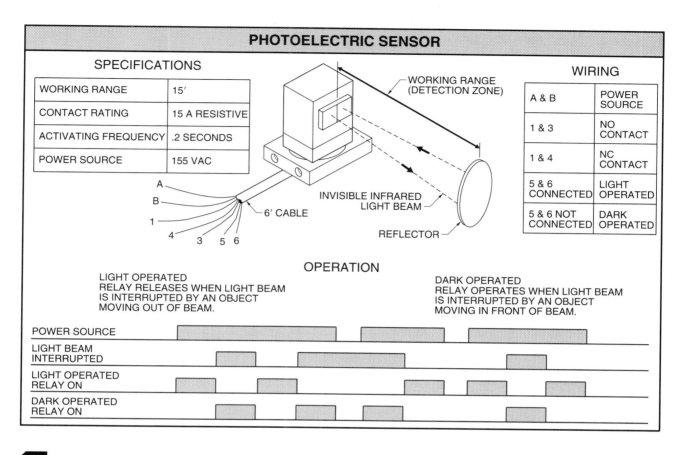

PHOTOELECTRIC SENSOR

SPECIFICATIONS

WORKING RANGE	15'
CONTACT RATING	15 A RESISTIVE
ACTIVATING FREQUENCY	.2 SECONDS
POWER SOURCE	155 VAC

A
B
1
4   3   5   6
6' CABLE

WORKING RANGE (DETECTION ZONE)

INVISIBLE INFRARED LIGHT BEAM

REFLECTOR

WIRING

A & B	POWER SOURCE
1 & 3	NO CONTACT
1 & 4	NC CONTACT
5 & 6 CONNECTED	LIGHT OPERATED
5 & 6 NOT CONNECTED	DARK OPERATED

OPERATION

LIGHT OPERATED
RELAY RELEASES WHEN LIGHT BEAM IS INTERRUPTED BY AN OBJECT MOVING OUT OF BEAM.

DARK OPERATED
RELAY OPERATES WHEN LIGHT BEAM IS INTERRUPTED BY AN OBJECT MOVING IN FRONT OF BEAM.

POWER SOURCE

LIGHT BEAM INTERRUPTED

LIGHT OPERATED RELAY ON

DARK OPERATED RELAY ON

Application — Proximity Sensor Installation

Proximity sensors have a sensing head that produces a radiated sensing field. This sensing field detects the target of the sensor. The sensing field must be kept clear of interference for proper operation. *Interference* is any object other than the object to be detected that is sensed by the sensor. Interference may come from objects close to the sensor or from other sensors. General clearances are required for most proximity sensors.

Flush Mounted Inductive and Capacitive Proximity Sensors

When flush mounting inductive and capacitive proximity sensors, a distance equal to or greater than twice the diameter of the sensors is required between sensors. If two sensors of different diameters are used, the diameter of the largest sensor is used for installation. **See Flush Mounted.**

For example, if two 8 mm inductive proximity sensors are flush mounted, at least 16 mm is required between the sensors.

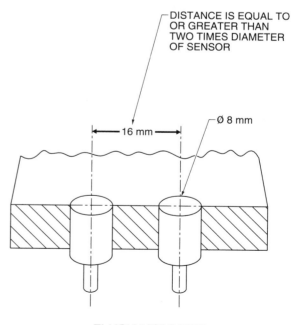

DISTANCE IS EQUAL TO OR GREATER THAN TWO TIMES DIAMETER OF SENSOR

16 mm

Ø 8 mm

FLUSH MOUNTED

Non-flush Mounted Inductive and Capacitive Proximity Sensors

When non-flush mounting inductive and capacitive proximity sensors, a distance of three times the diameter of the sensor is required within or next to a material that can be detected. Three times the diameter of the largest sensor is required when inductive and capacitive proximity sensors are installed next to each other. Spacing is measured from center to center of the sensors. When inductive and capacitive proximity sensors are mounted opposite each other, six times the rated sensing distance is required for proper operation. **See Non-flush Mounted.**

For example, if two 16 mm capacitive proximity sensors are non-flush mounted, at least 48 mm is required between the sensors.

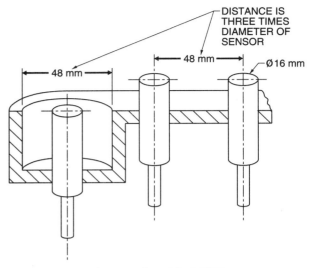

NON-FLUSH MOUNTED

Mounting Photoelectric Sensors

Photoelectric sensors transmit a beam of light. The beam of light detects the presence (or absence) of an object. Only part of the light beam is effective when detecting the object. The *effective light beam* is the area of light that travels directly from the transmitter to the receiver. If the object does not completely block the effective light beam, the object is not detected.

When mounting photoelectric sensors, the receiver is positioned to receive as much light as possible from the transmitter. Because more light is available at the receiver, greater operating distances are allowed and more power is available for the system to see through dirt in the air and on the transmitter and receiver lenses. The transmitter is mounted on the clean side of the detection zone because light scattered by dirt on the receiver lens affects the system less than light scattered by dirt on the transmitter lens.

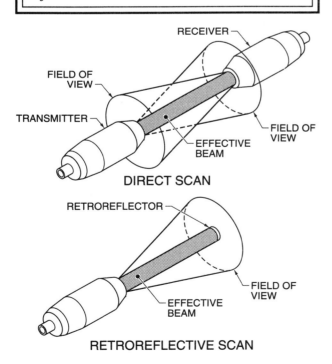

DIRECT SCAN

RETROREFLECTIVE SCAN

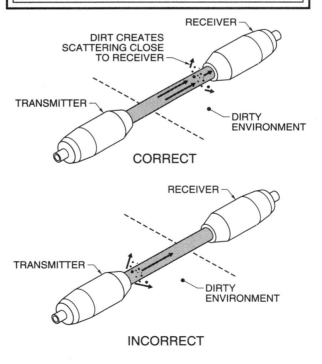

CORRECT

INCORRECT

Application — Determining Activating Frequency

Activating frequency is the limit to the number of pulses per second that can be detected by a photoelectric control in a time period. All photoelectric controls have an activating frequency. To determine the required activating frequency of a photoelectric application, apply the procedure:

Step 1. Determine the maximum speed of the objects to be detected. The speed of the objects is found by measuring the speed of the objects (in ft/min, or in./sec) and converting the speed to sec/in. **See Speed Conversions.**

For example, cartons on a conveyor travel at 21'/min. Twenty-one feet per minute equals .238 sec/in. (from Speed Conversions).

Step 2. Determine the dark input signal duration. *Dark input signal duration* is the time period when the photosensor is dark because the detected object is blocking the light beam. Dark input signal duration is found by multiplying the minimum dimension of the object to be detected (in in.) by the sec/in. value.

For example, if a 6″ × 3″ container has a .263 sec/in. value, the dark input signal duration is .789 seconds (3 × .263).

Step 3. Determine the light input signal duration. *Light input signal duration* is the time period when the photosensor is lit because no detectable object is in the light beam. Light input signal duration is found by multiplying the minimum distance between the objects to be detected by the sec/in. value.

For example, if the moving objects are spaced between 4″ and 15″ on a conveyor traveling at .200 sec/in., the light input signal duration is .8 seconds (4 × .200).

Step 4. Determine activating frequency. Activating frequency of a photoelectric control application is found by adding the dark input signal duration to the light input signal duration. This value is compared to the manufacturer's stated value. The activating frequency of a photoelectric control must be less than the activating frequency required for the application.

Example: Finding Activating Frequency

The minimum distance between objects in an application is 5″. The dimensions of the objects are 2″ × 2″. The maximum speed of travel is 40'/min. Find the activating frequency.

Step 1. Determine maximum speed of travel in sec/in.

Maximum speed of travel is .125 sec/in. (from Speed Conversions).

Step 2. Determine dark input signal duration.

Dark input signal duration is .25 seconds (2 × .125).

Step 3. Determine light input signal duration.

Light input signal duration is .625 seconds (5 × .125).

Step 4. Determine activating frequency.

Activating frequency is .875 seconds (.25 + .625). To work properly, the photoelectric control must have an activating frequency of .875 sec or less.

SPEED CONVERSIONS				
Ft/min	In./min	Ft/sec	In./sec	Sec/in.
1	12	.017	.2	5
3	36	.050	.6	1.666
5	60	.083	1.0	1.000
7	84	.116	1.4	.714
9	108	.150	1.8	.555
11	132	.183	2.2	.435
13	156	.216	2.6	.385
15	180	.249	3.0	.333
17	204	.282	3.4	.294
19	228	.315	3.8	.263
21	252	.349	4.2	.238
23	276	.382	4.6	.2172
25	300	.415	5	.200
40	480	.664	8	.125
60	720	.996	12	.0833
80	960	1.328	16	.0625
100	1200	1.66	20	.050
200	2400	3.32	40	.025
250	3000	4.15	50	.020
300	3600	4.98	60	.016
400	4800	6.64	80	.012
500	6000	8.30	100	.010
700	8400	11.62	140	.007
1250	15,000	20.75	250	.004
2500	30,000	41.5	500	.002

Application — Applying Photoelectric Sensors

Photoelectric sensors are used in many applications because they detect the presence or absence of any object without touching the object. Because of this no-touch feature, proximity sensors can be used in applications where mechanical limit switches cannot. The sensors can detect light objects, heavy objects, or untouchable objects. Untouchable objects include hot objects and freshly painted objects. **See Proximity Sensor Applications.**

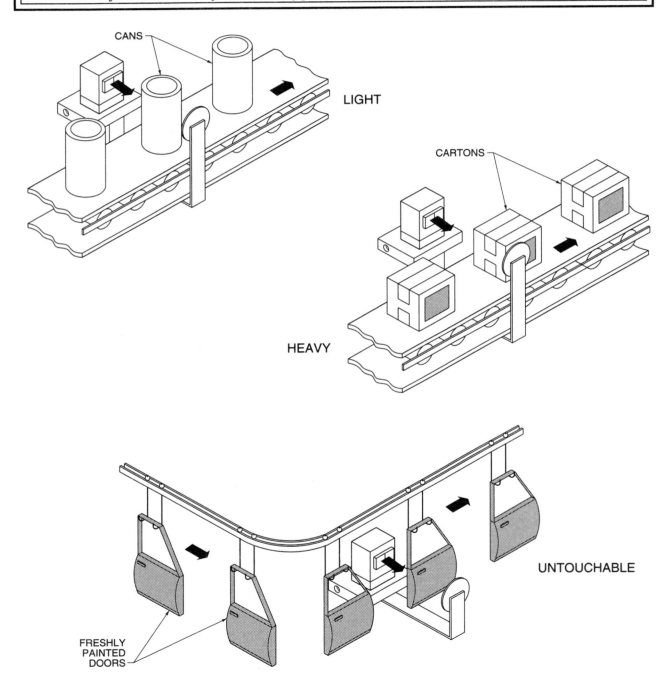

CANS

LIGHT

CARTONS

HEAVY

UNTOUCHABLE

FRESHLY
PAINTED
DOORS

Name _____   Date _____

⬤ Activity 13-1. Mounting Proximity Sensors

_____ **1.** The minimum distance required between the sensors for proper operation is _____ mm.

_____ **2.** The minimum distance required between the sensor and the surrounding material for proper operation is _____ mm.

_____ **3.** The minimum distance required between the sensors and the surrounding material for proper operation is _____ mm.

⬤ Activity 13-2. Determining Activating Frequency

Answer the questions using Speed Conversions on page 121.

_____ **1.** A photoelectric sensor detects 2″ × 2″ objects that are 5″ apart and travel at 60′/min.

_____ **A.** The dark input signal duration is _____ seconds.

_____ **B.** The light input signal duration is _____ seconds.

_____ **C.** The activating frequency of the application is _____ seconds.

    2. A photoelectric sensor detects .5″ square objects that are 2″ apart and travel at 200′/min.

_____   **A.** The dark input signal duration is _____ seconds.

_____   **B.** The light input signal duration is _____ seconds.

_____   **C.** The activating frequency of the application is _____ seconds.

    3. A photoelectric sensor detects 2″ square objects that are 3″ apart and travel at 1250′/min.

_____   **A.** The dark input signal duration is _____ seconds.

_____   **B.** The light input signal duration is _____ seconds.

_____   **C.** The activating frequency of the application is _____ seconds.

    4. A photoelectric sensor detects .25″ × .25″ objects that are .25″ apart and travel at 15′/min.

_____   **A.** The dark input signal duration is _____ seconds.

_____   **B.** The light input signal duration is _____ seconds.

_____   **C.** The activating frequency of the application is _____ seconds.

◯ Activity 13-3. Applying Photoelectric Sensors

Photoelectric sensors detect possible product jam-ups. Design the control circuit using a photoelectric sensor. Use a standard start/stop pushbutton station to control the conveyor motor. If the photoelectric sensor is blocked (product jam-up) for more than 5 seconds, the conveyor stops and an alarm sounds. Mark each wire number used from the photoelectric sensor. See Photoelectric Sensor on page 119.

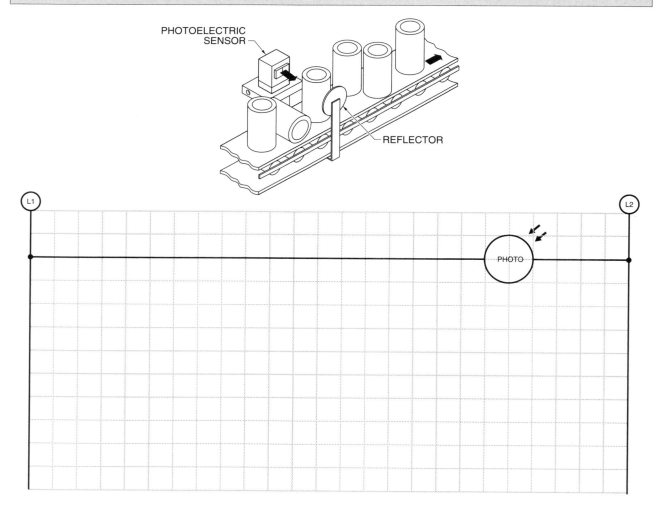

Activity 13-4.   Applying Photoelectric Sensors

Two photoelectric sensors control the merging of two conveyors. Design the control circuit using two photoelectric sensors. Use a standard start/stop pushbutton station to control Conveyor 1. Conveyor 2 runs when Conveyor 1 is running and when two boxes are not merging simultaneously. Design the circuit so both conveyor motors operate if no box is detected. If Photoelectric Sensor 1 detects a box and Photoelectric Sensor 2 does not, both conveyors continue to run. If Photoelectric Sensor 1 and 2 both detect a box, Conveyor 1 continues to run, and Conveyor 2 stops until Conveyor 1 is clear. Mark each wire number used from each photoelectric sensor. See Photoelectric Sensor on page 119.

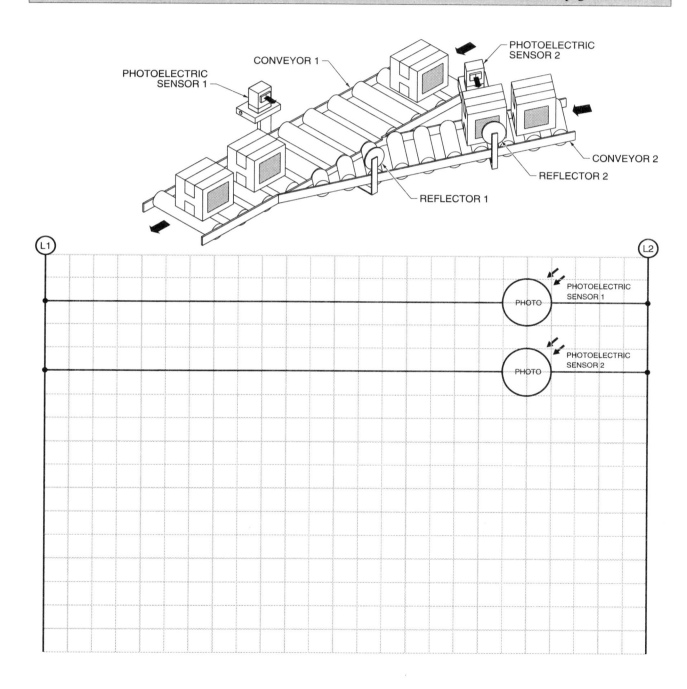

⬤ Activity 13-5. Applying Capacitive Proximity Sensors

Four capacitive proximity sensors detect bottles through a carton. Capacitive proximity sensors can detect plastic, glass, or metal containers inside the carton. Draw the control circuit using four capacitive proximity sensors and four DC-to-AC solid-state relays. In the DC circuit, each capacitive proximity sensor energizes one light and one solid-state relay when a bottle is present. In the AC circuit, the four solid-state relay contacts energize an operate delay relay. The relay is set for 5 seconds and controls the conveyor motor starter. When all four sensors detect a bottle, the conveyor motor runs for 5 seconds and automatically advances each carton after it is filled. If the conveyor does not advance, an operator checks the lights to see which bottle(s) have not dropped. If the carton advances, the circuit is automatically reset when the carton passes the four sensors. See Capacitive Sensor (PNP) on page 118. Mark the color of each wire used on the sensor.

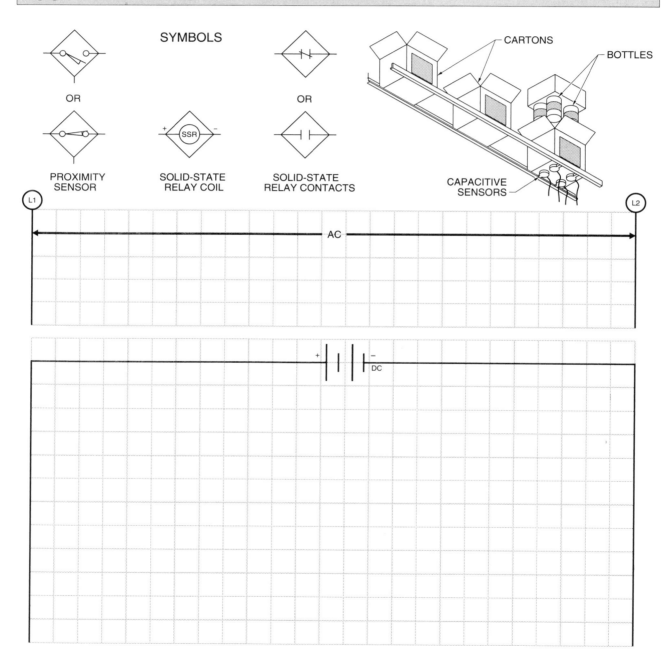

◯ Activity 13-6. Applying Inductive Proximity Sensors

A drill motor is used to drill a hole in a workpiece. Proximity Sensor 1 detects the presence of the workpiece. Proximity Sensor 2 automatically returns the drill. The movement of the drill motor is controlled by a hydraulic valve. The valve is controlled by two solenoids. Draw the control circuit using the two sensors and two DC-to-AC solid-state relays. In the DC circuit, each sensor controls a solid-state relay. In the AC circuit, a pushbutton energizes the advance solenoid if the workpiece is in place. The other sensor automatically returns the drill motor. See Inductive Sensor (PNP) on page 117. Mark the color of each wire used on the proximity sensor.

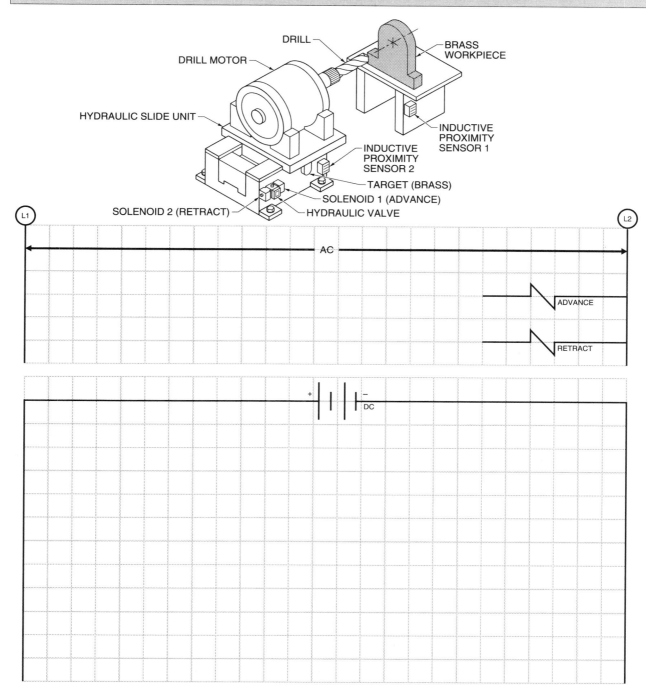

○ Activity 13-7. Applying Capacitive Proximity Sensors

A capacitive proximity sensor detects the presence of glass panels moving along a production line. Draw the control circuit using a capacitive proximity sensor and two DC-to-AC solid-state relays. In the DC circuit, the proximity sensor energizes two solid-state relays when the glass panels are present. In the AC circuit, a three-position selector switch controls a two-speed conveyor motor that moves the glass panels along the production line. When the selector switch is in the low position, the low-speed starter is energized. When the low-speed starter is energized, glass panels move along the production line at 10 second intervals. When the selector switch is in the high position, the high-speed starter is energized. When the high-speed starter is energized, glass panels move along the production line at 15 second intervals. When the selector switch is in the OFF position, no starter is energized.

Design the circuit so an alarm sounds when the proximity sensor does not detect a glass panel every 15 seconds in low speed. An alarm also sounds when the proximity sensor does not detect a glass panel every 10 seconds in high speed. See Capacitive Sensor (PNP) on page 118. Mark the color of each wire used on the proximity sensor.

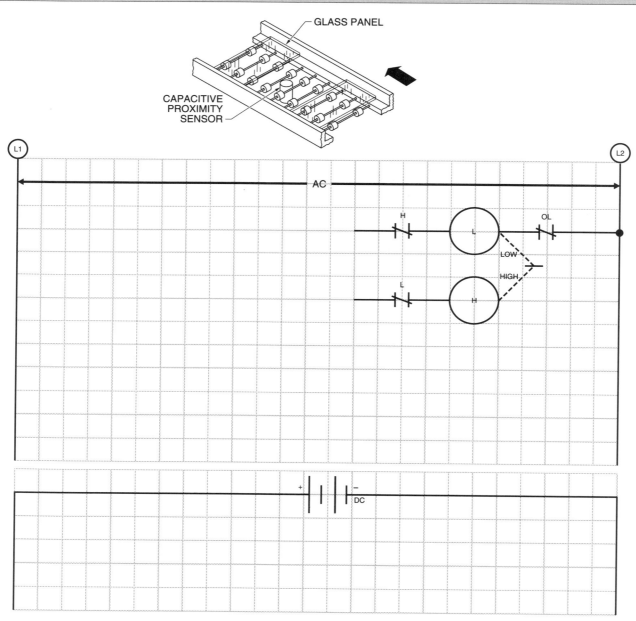

Application — System Input and Output Identification

Input Section

The input section of a programmable controller receives information from pushbuttons; temperature switches; pressure switches; overload contacts; and other manual, mechanical, or automatic inputs.

The inputs connected to the programmable controller are classified as digital or analog. *Digital inputs* are inputs that have only two positions, ON and OFF. Digital inputs include pushbuttons, switches, relay contacts, and mechanical limit switches. *Analog inputs* are inputs that change continuously over a range. Analog inputs include variable-voltage inputs, variable-current inputs, and variable-resistance inputs.

Output Section

The output section of a programmable controller delivers the output voltage to control alarms, lights, solenoids, motor starters, and other devices that produce work in the system. Like inputs, outputs connected to the programmable controller are digital or analog.

Digital outputs include lights, motor starters, alarms, solenoids, and contactors. Analog outputs include variable-output voltages, variable-current output, or variable-resistance output. The inputs and outputs must be identified when troubleshooting, wiring, or designing a system. **See Programmable Controller System.**

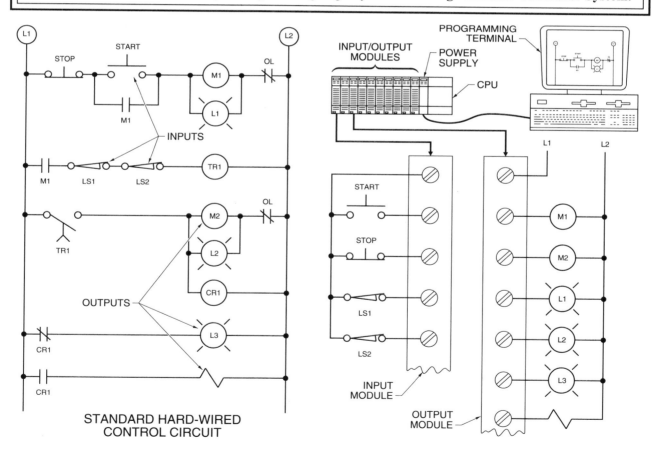

STANDARD HARD-WIRED
CONTROL CIRCUIT

Application — System Input and Output Connections

When connecting an input to a programmable controller, one side of the input is connected to an assigned input terminal and the other side is connected to a common terminal. The assigned input terminal is usually marked IN 1, IN 2, IN 12, IN 125, etc. The common terminal is usually marked COM or C. When connecting an output to a programmable controller, one side of the output is connected to an assigned output terminal, and the other side is connected to a common power line. The assigned output terminal is usually marked OUT 1, OUT 2, OUT 12, OUT 125, etc. The power line connections are usually marked as Ll/L2, +/-, +V/-V, VAC IN, or VDC IN. **See Programmable Controller Connections.**

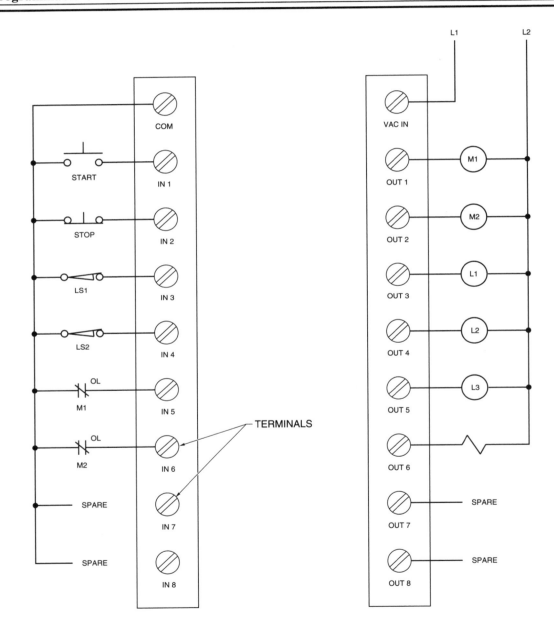

INPUT MODULE   OUTPUT MODULE

◼ Application — Alarm Output Connection

Programmable controllers include alarm output contacts that are activated if the reserve battery is low or other problems occur in the programmable controller. If the reserve battery is low and a power failure occurs, the program can be lost. When power returns, the programmable controller does not operate properly until it is reprogrammed. This can cause problems in the system. **See Programmable Controller Alarm.**

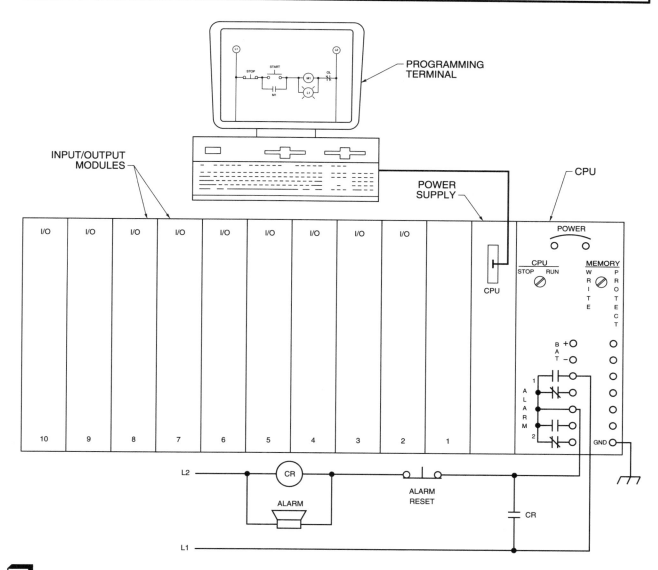

◼ Application — Applying a Programmable Controller

A programmable controller is used to control machine or process operations with a stored program that controls the outputs based on the status of the inputs. No wiring change of the inputs and outputs is required when the machine function or process is changed. The logic of the circuit is changed through the programming terminal. When a programmable controller is applied to a control circuit, the inputs and outputs are identified and connected to the input or output section. **See Motor Control Circuit.**

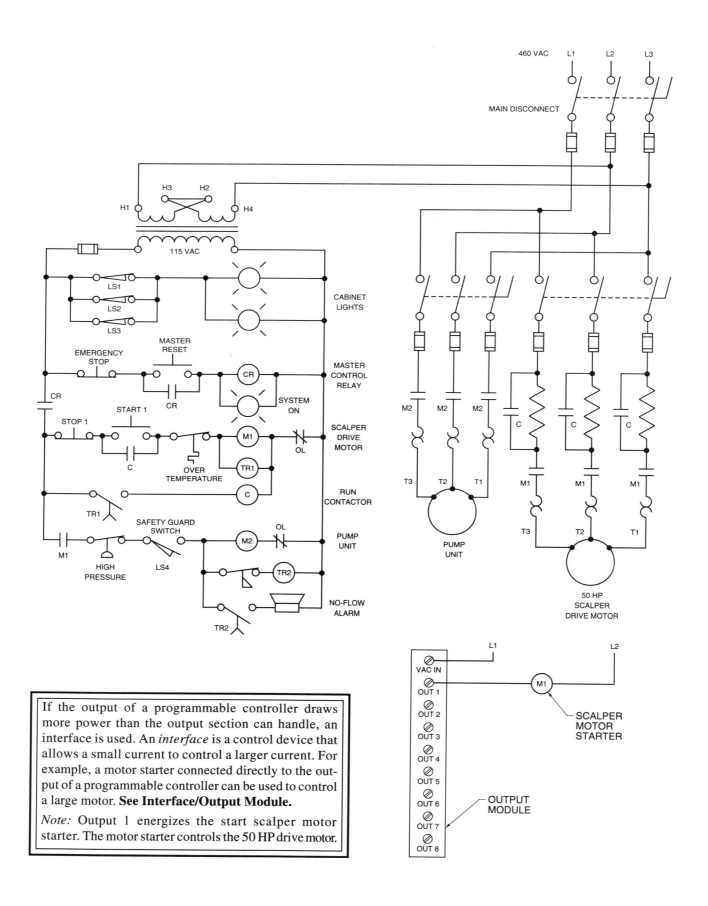

If the output of a programmable controller draws more power than the output section can handle, an interface is used. An *interface* is a control device that allows a small current to control a larger current. For example, a motor starter connected directly to the output of a programmable controller can be used to control a large motor. **See Interface/Output Module.**

Note: Output 1 energizes the start scalper motor starter. The motor starter controls the 50 HP drive motor.

Name _____   Date _____

Activity 14-1.   System Input and Output Identification

1. Identify the type (input or output), name (limit switch or temperature switch), and function (starts scalper motor) of the inputs and outputs for Motor Control Circuit.

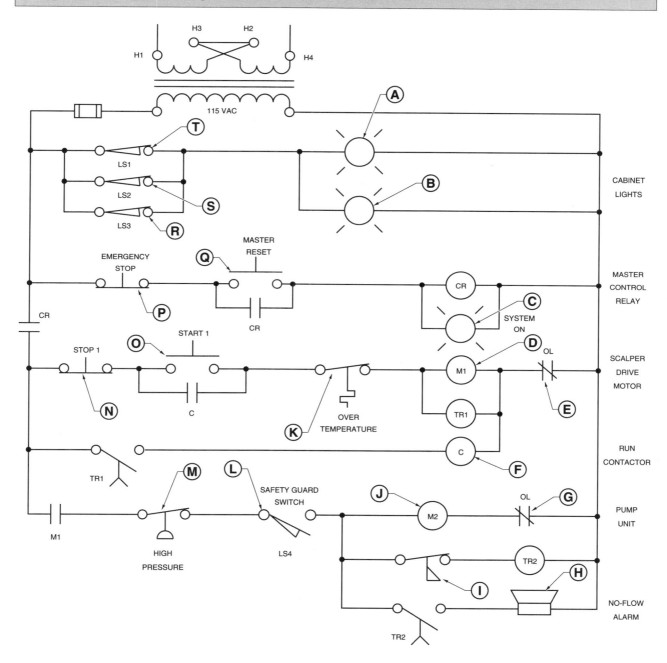

	Type	Name	Function
	MOTOR CONTROL CIRCUIT		
A	Output	Light	Light cabinet interior
B			
C			
D			
E			
F			
G			
H			
I			
J			
K			
L			
M			
N			
O			
P			
Q			
R			
S			
T			

⚪ **Activity 14-2.   System Input and Output Connections**

1. Connect the inputs to the input modules in the order they are listed on page 134 . Mark any extras as spares.

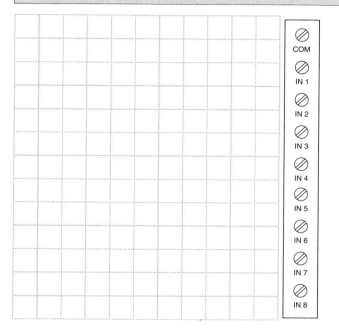

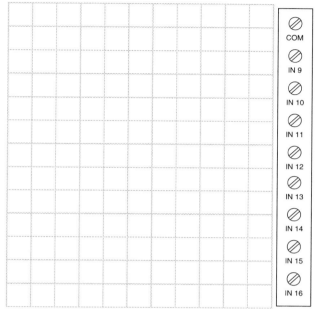

2. Connect the outputs to the output module in the order they are listed on page 134 . Mark any extras as spares.

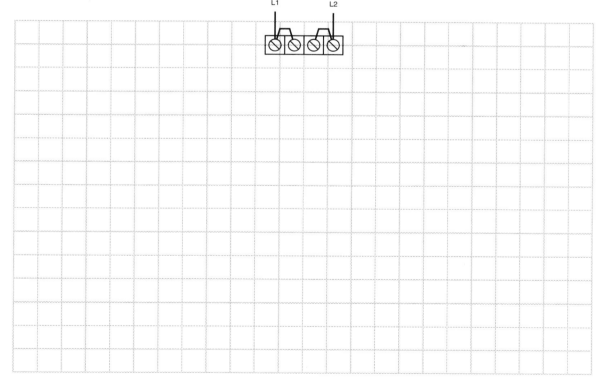

◯ Activity 14-3.   Alarm Output Connection

Nuisance tripping can occur when the alarm contacts on a programmable controller close for only a few seconds. To prevent nuisance tripping, a timer is added to the circuit to delay the activation of the alarm.

Draw the line diagram of the Programmable Controller Alarm on page 131 and add an operate delay relay to prevent the alarm from tripping unless the alarm contacts are closed for more than 3 seconds.

APPLICATIONS

Application — Primary Resistor Reduced-voltage Starting

Primary resistor reduced-voltage starting is a motor starting method that has resistors connected in series with a motor to reduce the voltage applied to the motor when starting. After a time period, full line voltage is applied to the motor. Primary resistor reduced-voltage starting provides a smooth acceleration through a simple, economical circuit. Primary resistor reduced-voltage starting applications include geared or belted drives where the sudden application of the full-voltage torque must be avoided. **See Primary Resistor Reduced-voltage Starting.**

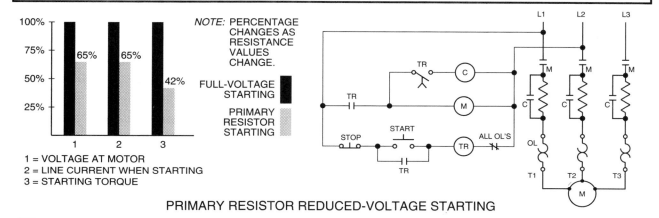

PRIMARY RESISTOR REDUCED-VOLTAGE STARTING

Application — Part-winding Reduced-voltage Starting

Part-winding reduced-voltage starting is a motor starting method that applies current to one-half of the motor windings when starting. After a time period, the starter applies current to all motor windings. Part-winding reduced-voltage starting is economical because it requires less components and space than other reduced-voltage starting methods. Part-winding reduced-voltage starting applications include low-inertia loads, such as commercial air conditioning compressors, pumps, fans, and blowers or locations where local power companies place limitations on the amount of inrush current. Part-winding reduced-voltage starting requires a nine-lead, 3φ motor. **See Part-winding Reduced-voltage Starting.**

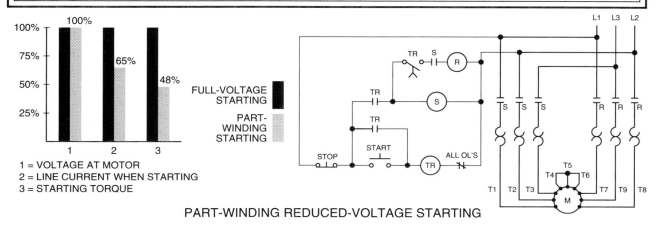

PART-WINDING REDUCED-VOLTAGE STARTING

Application — Autotransformer Reduced-voltage Starting

Autotransformer

Autotransformer reduced-voltage starting is a motor starting method that reduces the applied motor voltage to 50%, 65%, or 80% of the line voltage when starting. This is done by placing a transformer coil in series with the motor for a time period. After the time period, the motor is connected to full line voltage. Autotransformer reduced-voltage starting is used for starting blower, compressor, conveyor, and pump motors above 10 HP.

Autotransformer reduced-voltage starting provides the highest possible starting torque per ampere of line current. However, because autotransformers are required, installation cost is higher than other reduced-voltage starting methods. Autotransformer reduced-voltage starting is used with any 3ϕ motor. **See Autotransformer Reduced-voltage Starting.**

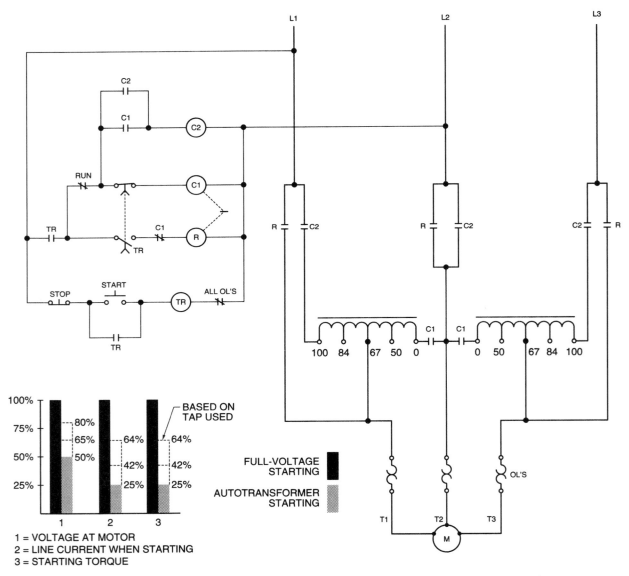

AUTOTRANSFORMER REDUCED-VOLTAGE STARTING

Application — Wye/delta Reduced-voltage Starting

Wye/delta

Wye/delta reduced-voltage starting is a motor starting method that has the motor connected as a wye motor when starting. This arrangement reduces the coil voltage of the motor to about 58% of line voltage. After a time period, the motor is connected as a delta motor. Wye/delta reduced-voltage starting is used where the power supply is inadequate to provide full starting current without an objectionable voltage drop or where low starting torque is required. Wye/delta reduced-voltage starting applications include fans, compressors, and conveyors that have long acceleration times or frequent starts. Wye/delta reduced-voltage starting is only used with six-lead, 3φ motors. **See Wye/delta Reduced-voltage Starting.**

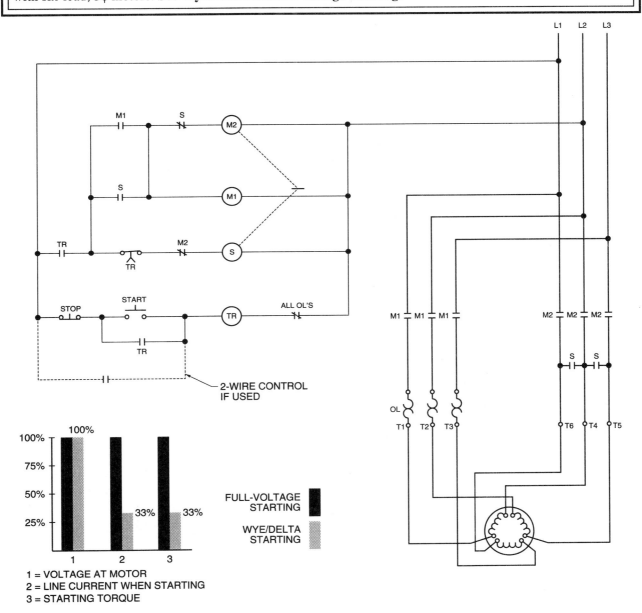

WYE/DELTA REDUCED-VOLTAGE STARTING

Application — Closed Transition Reduced-voltage Starting

Closed transition reduced-voltage starting is a method of motor starting in which the motor is not disconnected from the power source during the transition from start to run. There is a brief time period between the point when the motor is disconnected from the reduced voltage and the point when it is reconnected to full line voltage in autotransformer or wye/delta reduced-voltage starting. This open transition from start to run causes a flow of high transient currents when the motor is reconnected to full line voltage. To eliminate this problem, closed transition starting is added. **See Closed Transition Reduced-voltage Starting.**

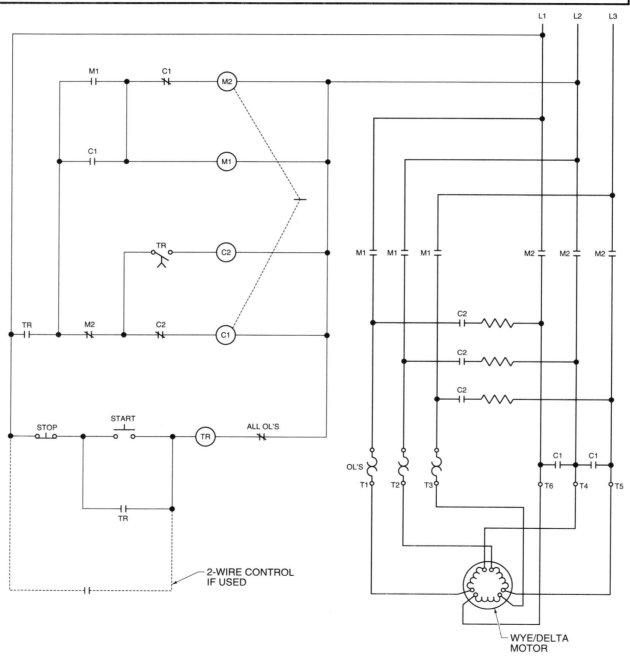

CLOSED TRANSITION REDUCED-VOLTAGE STARTING

ACTIVITIES

Name _____   Date _____

◑ Activity 15-1. Primary Resistor Reduced-voltage Starting

Connect the power circuit from the transformer bank to the motor for high-voltage, reduced-voltage starting. Make all power circuit connections at the terminal screws. Connect the control transformer for 440 V to 110 V. See Dual-voltage Transformer on page 53. Draw a control circuit that includes a two-position selector switch and a pressure switch. When the selector switch is in the automatic position, the motor is controlled by the pressure switch. The motor energizes at low pressure. When the selector switch is in the manual position, the motor is controlled by a start/stop pushbutton station. Include a light that indicates the motor is running at full voltage.

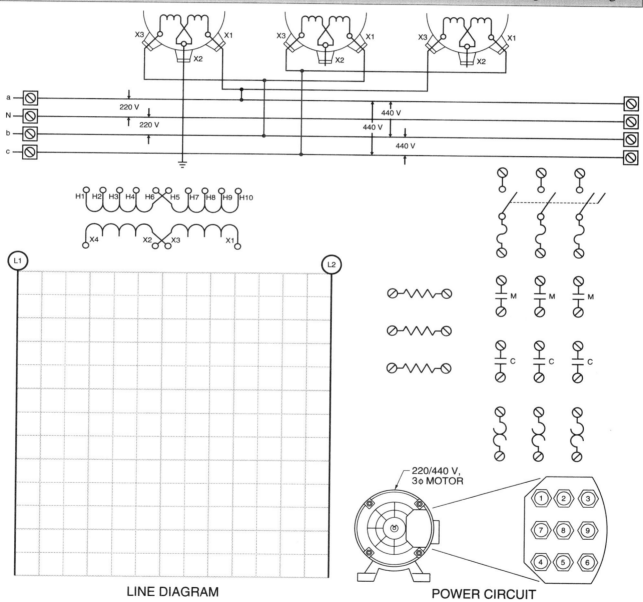

LINE DIAGRAM

POWER CIRCUIT

141

◖ Activity 15-2. Part-winding Reduced-voltage Starting

Connect the power circuit from the transformer bank to the motor for low-voltage, reduced-voltage starting. Make all power circuit connections at the terminal screws. Connect the control transformer for 208 V to 110 V. In Part-winding Reduced-voltage Starting Circuit on page 137, the timer has both instantaneous and time-delay contacts. Modify the control circuit to include a timer that has only time-delay contacts. See Models A and B Operate Delay Relays on page 59. Do not change the operation of the circuit.

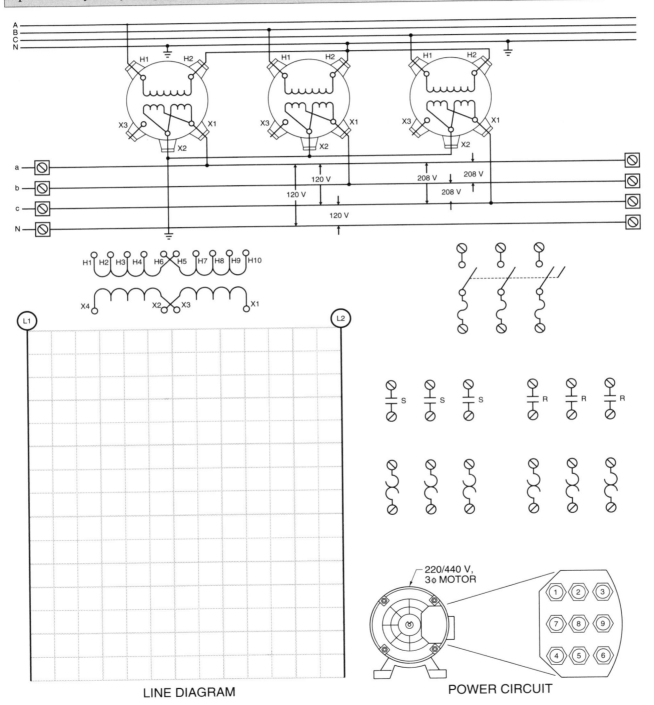

LINE DIAGRAM

POWER CIRCUIT

◯ Activity 15-3. Wye/delta Reduced-voltage Starting

Connect the motor for reduced-voltage starting. Make all power circuit connections at the terminal screws. Draw a control circuit that includes a two-position selector switch and a temperature switch. When the selector switch is in the automatic position, the motor is controlled by the temperature switch. The motor energizes at high temperature. When the selector switch is in the manual position, the motor is controlled by a start/stop pushbutton station. Modify the circuit to include a timer that has only time-delay contacts. See Models A and B Operate Delay Relays on page 59.

LINE DIAGRAM

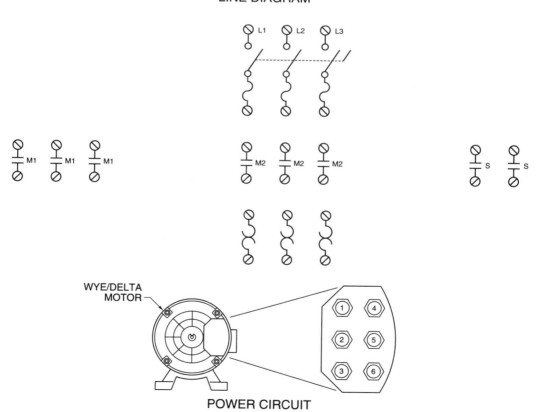

WYE/DELTA MOTOR

POWER CIRCUIT

⃝ Activity 15-4.   Closed Transition Reduced-voltage Starting

Connect the motor for closed transition reduced-voltage starting. Make all power circuit connections at the terminal screws.

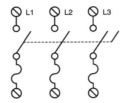

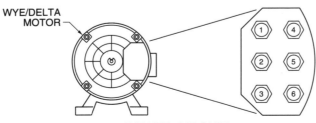

WYE/DELTA MOTOR

POWER CIRCUIT

Application — One-direction Motor Plugging

Plugging is a method of motor braking in which motor connections are reversed so the motor develops a countertorque that acts as a braking force. Plugging a motor to a rapid stop is done with a plugging switch. The plugging switch prevents the motor from reversing direction when the motor comes to a stop. The plugging switch automatically interrupts the reversing braking power as the motor approaches zero speed. The speed at which the plugging switch contacts operate is adjusted to avoid coasting or reverse rotation of the motor. Plugging is used for emergency stops to protect personnel when a motor must be stopped quickly. Plugging is also used to protect machines and machine tools. When using plugging as a stopping method, the following factors should be considered:

1. The driven machine, belts, chains, or couplings may not be able to withstand the forces created by plugging.

2. When plugging a motor, approximately six to eight times the normal full-load current of the motor is drawn. The power supply and control equipment must be sized to withstand this overload.

3. The motor may not be sized to withstand frequent plugging. A larger motor or a motor with a higher duty rating may be used. **See One-direction Motor Plugging.**

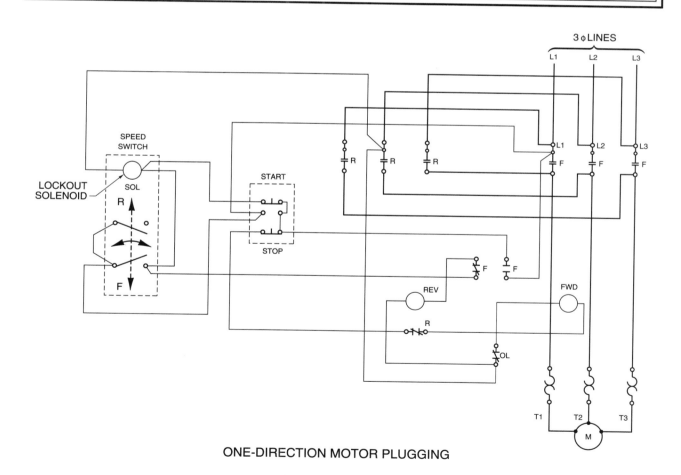

ONE-DIRECTION MOTOR PLUGGING

◻ Application — Two-direction Motor Plugging

In some applications, a motor must be brought to a rapid stop in either direction. Applications include reversing control circuits used around personnel. When a motor must be plugged to a stop in either direction, a two-direction plugging switch is used. **See Two-direction Motor Plugging.**

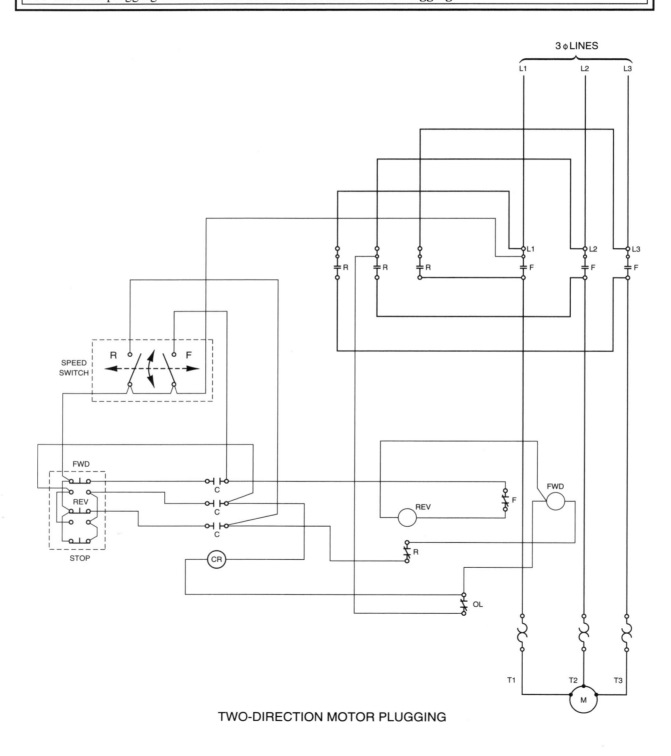

TWO-DIRECTION MOTOR PLUGGING

▢ Application — Two-speed Separate Winding Motors

The speed of an AC motor is determined by the frequency of the power supply or the number of individual poles. The individual poles are determined by how the motor windings are connected. The speed of a motor decreases as the number of poles increases. The speed of a motor increases as the number of poles decreases.

To change the speed of a motor, the motor must have separate windings. Each winding has a different number of individual poles. When power is applied to different windings, the motor speed changes. In a two-speed motor circuit, the motor can be started in either high or low speed. The stop button must be pressed before changing speeds. **See Two-speed Separate Winding Motor.**

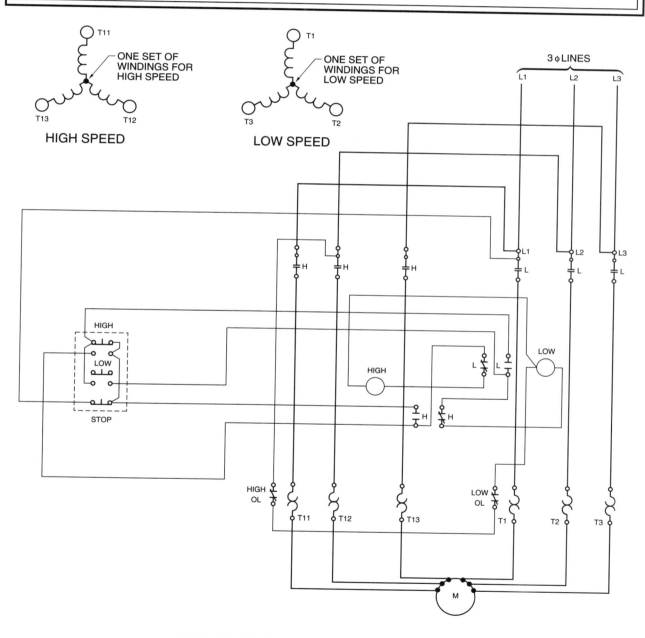

TWO-SPEED SEPARATE WINDING MOTOR

Application — Two-speed Consequent Pole Motors

Motor speed can be changed with consequent pole motors. *Consequent pole motors* are motors that have one winding connected and reconnected so that it has half or twice the original number of poles. As the number of poles on the winding changes, the speed of the motor changes.

In a two-speed consequent pole motor circuit, the motor can be started in high or low speed. The stop button must be pressed before changing speeds. **See Two-speed Consequent Pole Motor.**

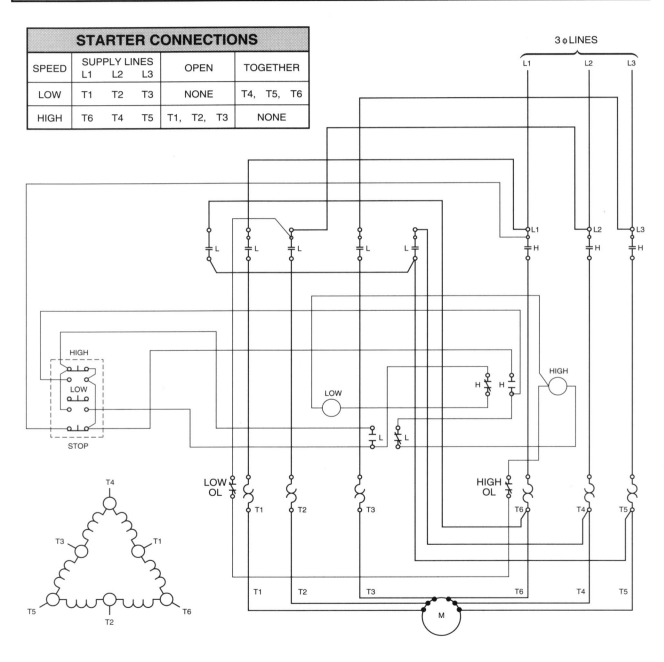

STARTER CONNECTIONS				
SPEED	SUPPLY LINES L1 L2 L3		OPEN	TOGETHER
LOW	T1	T2 T3	NONE	T4, T5, T6
HIGH	T6	T4 T5	T1, T2, T3	NONE

TWO-SPEED CONSEQUENT POLE MOTOR

ACTIVITIES

Name _____   Date _____

⬤ Activity 16-1.   One-direction Motor Plugging

1. Draw the line diagram of One-direction Motor Plugging on page 145.

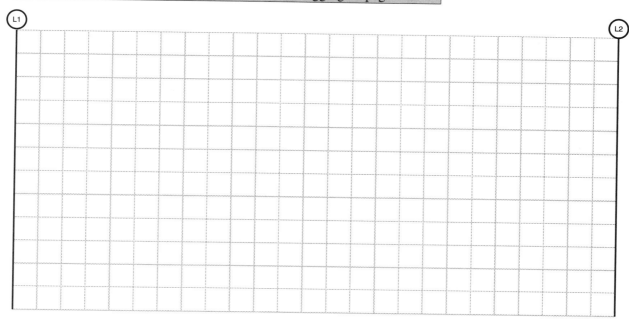

2. Redraw the line diagram and add a second stop pushbutton and a second start pushbutton.

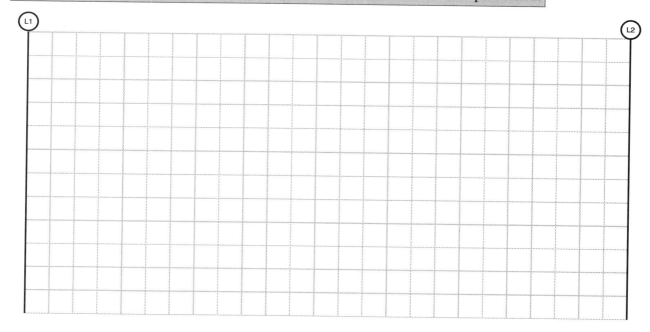

Activity 16-2. Two-direction Motor Plugging

1. Draw the line diagram of Two-direction Motor Plugging on page 146.

2. Redraw the line diagram and add a second forward pushbutton and a second reverse pushbutton. Include a yellow light to indicate that the motor is running in the forward direction and a red light to indicate the motor is running in the reverse direction.

Activity 16-3.   Two-speed Separate Winding Motors

1. Draw the line diagram of Two-speed Separate Winding Motor on page 147. *Note:* The stop pushbutton must be pressed before changing speeds.

2. Redraw the line diagram and add a red light to indicate the motor is running in high speed. Add a green light to indicate the motor is running in low speed. Include a temperature switch that automatically places the motor in high speed when a setpoint temperature is reached. Do not remove the high-speed pushbutton. The high-speed pushbutton is used to manually place the motor in high speed.

◖ Activity 16-4. Two-speed Consequent Pole Motors

1. Draw the line diagram of Two-speed Consequent Pole Motor on page 148.

2. Redraw the line diagram and replace the low- and high-speed pushbuttons with a three-position selector switch (low, OFF, high) and a pressure switch. If the selector switch is in the low or high position, the pressure switch automatically energizes the motor at a setpoint pressure. If the selector switch is in the OFF position, the motor does not operate.

Application — Power Circuit Troubleshooting

The *power circuit* is the part of a circuit that connects the loads to the main power lines. The loads are the devices that convert electrical energy to mechanical energy, heat, light, or sound. *Troubleshooting* is the systematic elimination of the various parts of a system to locate a malfunctioning part. Troubleshooting a power circuit is a matter of determining the point in the system at which power is lost. This point may be at the load, at the primary substation, or at any point in between. **See Power Circuit Distribution System.**

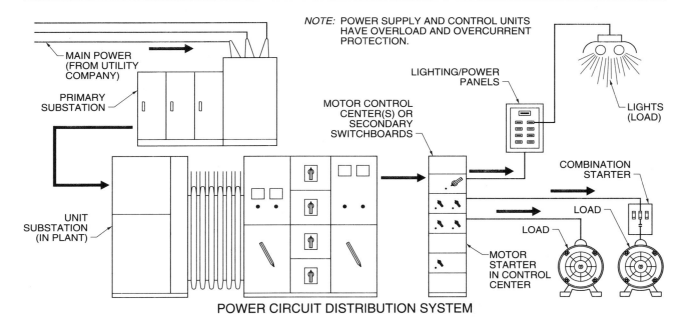

POWER CIRCUIT DISTRIBUTION SYSTEM

To troubleshoot a power circuit, apply the procedure:

Step 1. Check fuses, circuit breakers, and overload contacts.

All power circuits are protected from overcurrents at several points in the system by fuses and circuit breakers. For individual load problems, start with the fuses, circuit breakers, and overloads (on motor starters) closest to the load, and then work back through the system. For multiload problems, start with the motor control center, secondary switchboard, lighting panel, or power panel that feeds the loads.

Step 2. Check control circuit.

When energizing small loads, the control circuit may directly switch the power to the load. When energizing large loads, the control circuit uses an interface such as a contactor or motor starter to energize the load. Check to make sure the control circuit is delivering power to the load. If the control circuit is delivering power as required, the problem is not in the control circuit. If the control circuit is not delivering power as required, the problem is in the control circuit.

Step 3. Check load.

If power is delivered to the load but the load does not work, the problem is in the load. If the power delivered to the load is correct, the load needs replacement or service.

Application — Control Circuit Troubleshooting

The *control circuit* is the part of the circuit that determines when and how the loads are turned ON and OFF. Troubleshooting the control circuit is a matter of finding the point where the control power is lost. The point where power is lost usually indicates a malfunctioning switch, interface, or load in the control circuit. **See Control Circuit.**

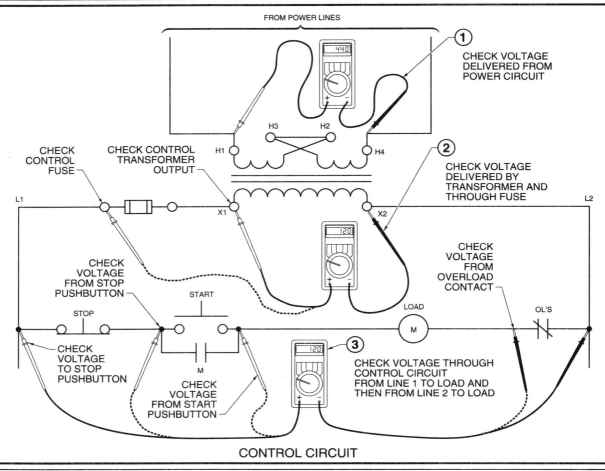

CONTROL CIRCUIT

To troubleshoot a control circuit, apply the procedure:

Step 1. Check voltage delivered from power circuit.

All control circuits receive voltage from a power circuit. Check the voltage coming from the power circuit to make sure voltage is present and is at the correct level.

Step 2. Check voltage delivered through control transformer.

Most control circuits operate at a lower voltage than the power circuit. A control transformer is used to reduce the voltage to the control circuit. Most control transformers have a fuse on the secondary side or a fuse is added on primary side. Check that the correct voltage is delivered by the transformer and is delivered through the fuse.

Step 3. Check voltage through control circuit.

Check the voltage through the control circuit by starting at Line 1 and Line 2. If the correct voltage is between Line 1 and Line 2, move one lead of the voltmeter through the control circuit. Start with Line 1 and move to the load in the control circuit. When a device does not pass the voltage as required, replace or service that device.

◯ Activity 17-1. Troubleshooting a Motor Control Circuit

> Answer the questions using Motor Control Circuit.

_____ **1.** The correct placement of Meter A so the meter checks the fuse in Line 3 is position _____.

_____ **2.** If the circuit is working correctly, can Meter B ever read 230 V at the motor when Meter C is reading 0 V?

_____ **3.** If the circuit is working correctly, can Meter C ever read 230 V at the motor when Meter B is reading 0 V?

_____ **4.** If Meter D reads 230 V in position 1 and 230 V in position 2, _____ Fuse is bad.

_____ **5.** If Meter E reads 230 V in position 1 and 0 V in position 2, Fuse _____ is bad.

_____ **6.** Meter F checks the fuse in Line _____.

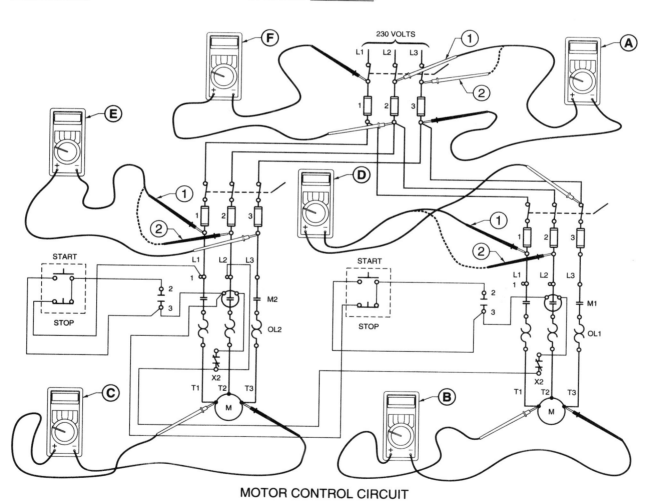

MOTOR CONTROL CIRCUIT

Activity 17-2.   Troubleshooting a Heating Circuit

Connect Meter A so the meter reads the primary voltage at the control transformer. Connect Meter B so the meter reads the voltage at the heating contactor. Connect Meter C so the meter checks the voltage delivered from the temperature switch when the switch is closed. Connect Meter D so the meter reads the voltage applied to the input of the temperature switch.

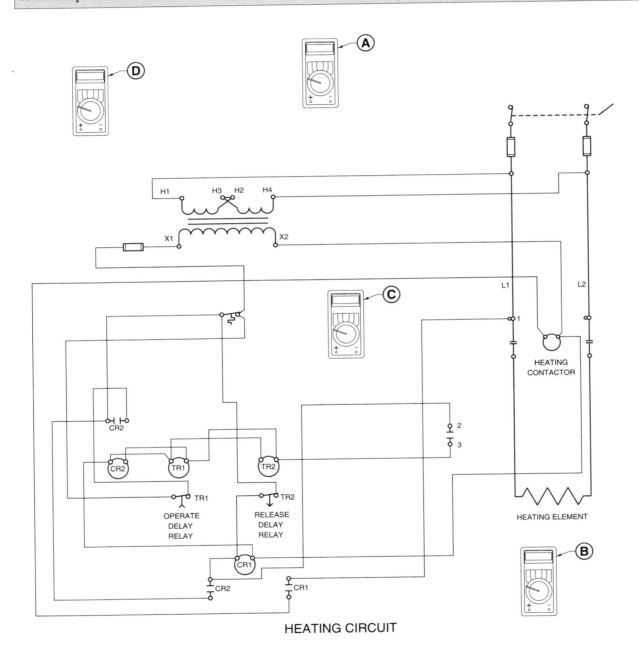

HEATING CIRCUIT

Activity 17-3. Troubleshooting a Reversing Motor Circuit

Determine the malfunctioning component using Reversing Motor Circuit and meter readings. *Note:* The machine operator reports that the motor operates in the forward direction but not in the reverse direction. Meter A reads 36 V when the forward pushbutton is pressed. Meter B reads 0 V when the forward or reverse pushbutton is pressed. Meter C reads 36 V when the reverse pushbutton is pressed. Meter D reads 0 V when the reverse pushbutton is pressed.

_____ **1.** The problem is in _____.

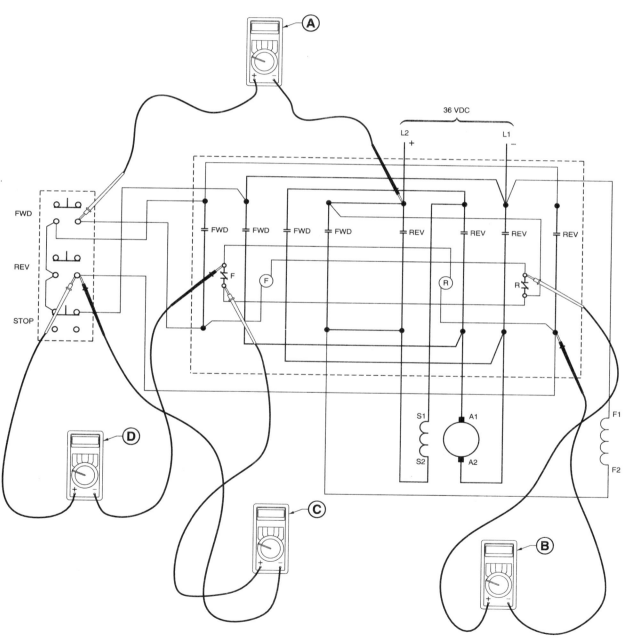

REVERSING MOTOR CIRCUIT

Activity 17-4. Troubleshooting a Dual-motor Control Circuit.

Connect Meter A so the meter checks Fuse 2. Connect Meter B so the meter reads the voltage at Motor Starter 2. Connect Meter C so the meter reads the voltage out of the start pushbutton when the start pushbutton is pressed.

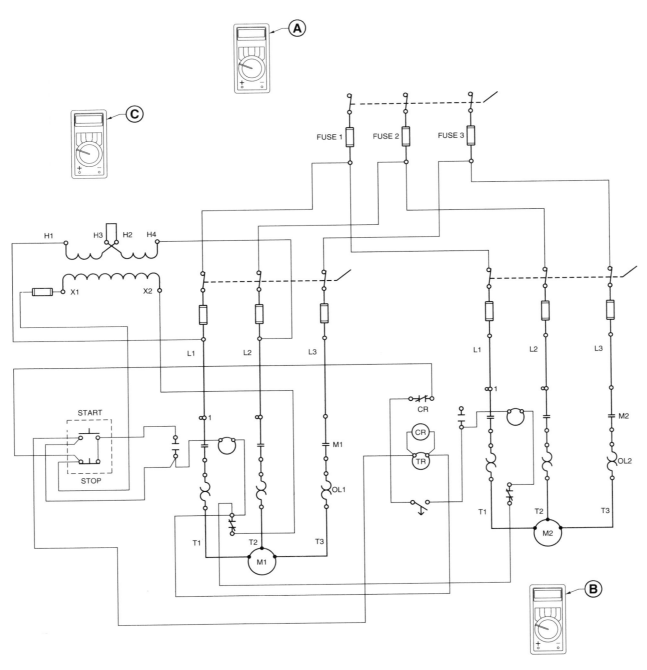

DUAL-MOTOR CONTROL CIRCUIT

◐ Activity 17-5. Troubleshooting a Selector Switch Circuit

Answer the questions using Selector Switch Circuit. *Note:* The motor does not operate in the automatic or OFF position. The motor does operate in the hand position.

_____ 1. The _____ is the most likely cause of the malfunction.
_____ 2. Could the problem be in the overloads?
_____ 3. Could the problem be in the motor starter coil?
_____ 4. Could the problem be in the selector switch?
_____ 5. Could the problem be in the pressure switch?
_____ 6. Could the problem be in the motor?

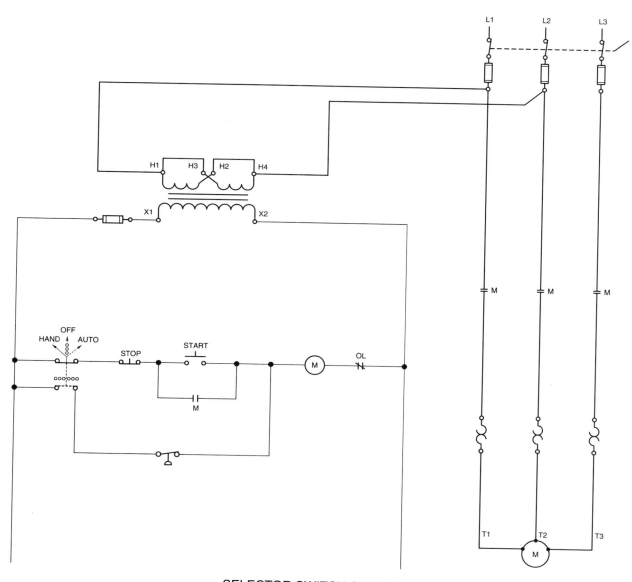

SELECTOR SWITCH CIRCUIT

◐ Activity 17-6.   Troubleshooting a Start/Stop Circuit

Answer the questions using Start/Stop Circuit. *Note:* The motor operates when the motor starter power contacts are manually closed by pressing them down at the motor starter. The motor does not operate when the start pushbutton is pressed or when a fused jumper test wire is connected from points 1 to 3. The meter reads 120 V.

_____   **1.** The _____ is the most likely cause of the malfunction.
_____   **2.** Could the problem be in the overloads?
_____   **3.** Could the problem be in the motor starter coil?
_____   **4.** Could the problem be in the stop pushbutton?
_____   **5.** Could the problem be in the start pushbutton?
_____   **6.** Could the problem be in the motor?

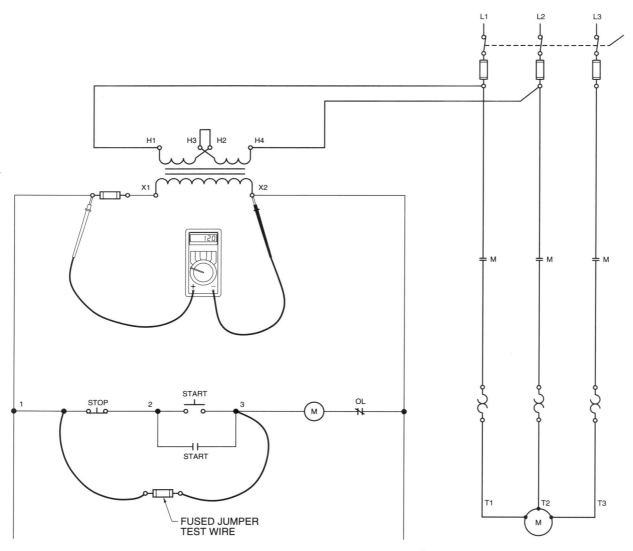

START/STOP CIRCUIT

◯ Activity 17-7. Troubleshooting a Control Transformer Circuit

Answer the questions using Control Transformer Circuit. *Note:* The motor operates when the motor starter power contacts are manually closed by pressing them down at the motor starter. The motor does not operate when the start pushbutton is pressed or when a fused jumper test wire is connected from points 1 to 3. A voltmeter connected at points X1 and X2 indicates proper voltage. A voltmeter connected at points 3 and X2 indicates no voltage at any time.

_____ **1.** The _____ is the most likely cause of the malfunction.
_____ **2.** Could the problem be in the overloads?
_____ **3.** Could the problem be in the motor starter coil?
_____ **4.** Could the problem be in the stop pushbutton?
_____ **5.** Could the problem be in the start pushbutton?
_____ **6.** Could the problem be in the transformer?
_____ **7.** Could the problem be in the control circuit fuse?

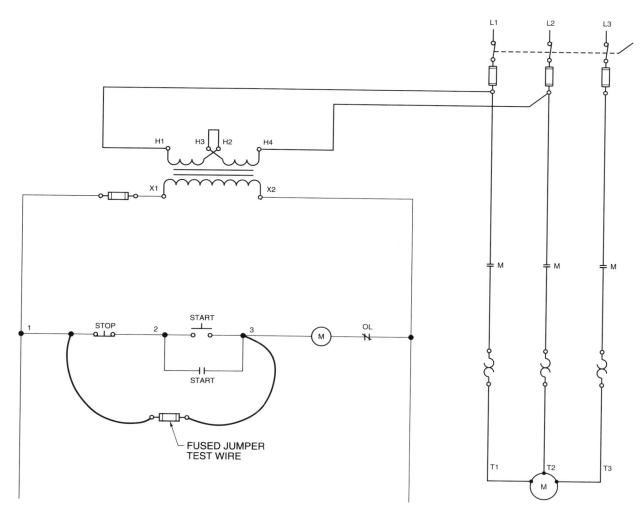

CONTROL TRANSFORMER CIRCUIT

◯ Activity 17-8. Troubleshooting a Primary Resistor Reduced-voltage Starting Circuit

Answer the questions using Primary Resistor Reduced-voltage Starting Circuit. *Note:* The motor operates when the start pushbutton is pressed, but does not seem to accelerate to full power. The machine operator reports that there is a burning smell coming from the control cabinet. The resistors get hot after the motor is started, even after several minutes. A voltage test taken at the motor terminals indicates lower voltage than at the incoming power lines.

_____ 1. A(n) _____ is the most likely cause of the malfunction.
_____ 2. Could the problem be in the pushbutton station?
_____ 3. Could the problem be in the motor starter?
_____ 4. Could the problem be in the contactor?
_____ 5. Could the problem be in the resistors?
_____ 6. Could the problem be in the timer?
_____ 7. Could the problem be in the overload contact?

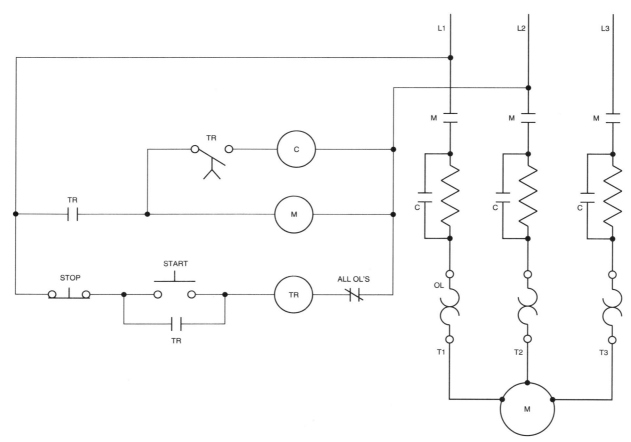

PRIMARY RESISTOR REDUCED-VOLTAGE STARTING CIRCUIT

Activity 17-9. Troubleshooting a Two-speed Separate Winding Circuit

Identify the malfunctioning component using Two-speed Separate Winding Circuit and the meter readings. *Note:* The machine operator reports that the motor operates in the low speed but not in the high speed. Meter A reads 440 V when the low pushbutton is pressed. Meter B reads 440 V when the low pushbutton is pressed. Meter C reads 0 V when the high pushbutton is pressed. Meter D reads 440 V when the high pushbutton is pressed.

_____ **1.** The component that has failed is the _____.

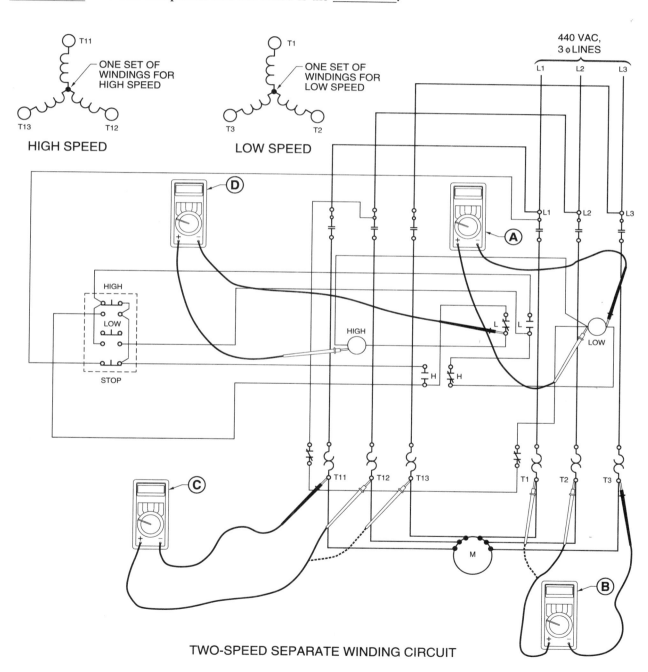

TWO-SPEED SEPARATE WINDING CIRCUIT

◑ Activity 17-10. Troubleshooting a Two-direction Motor Plugging Circuit

Connect Meter A to monitor the malfunctioning component when the motor is turned OFF in the forward direction. Connect Meter B to monitor the malfunctioning component when the motor is turned OFF in the reverse direction. *Note:* The machine operator reports that the motor does not come to a rapid stop in either direction.

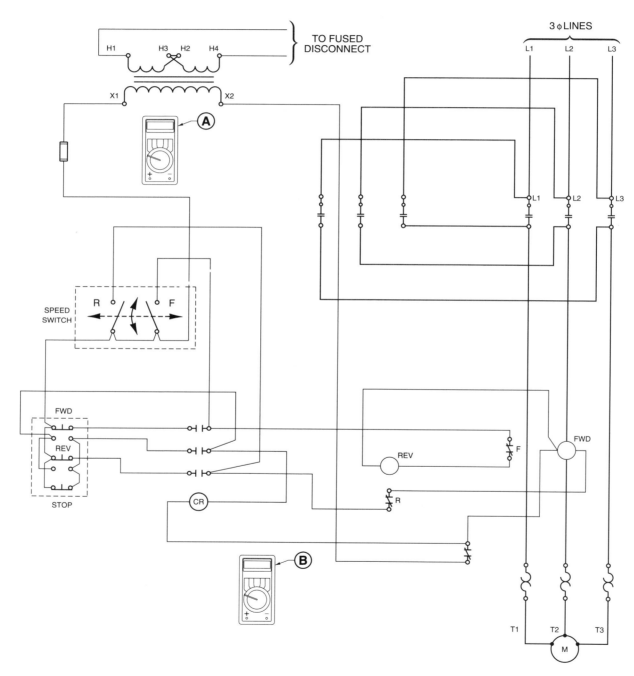

TWO-DIRECTION MOTOR PLUGGING CIRCUIT

COMMON PREFIXES		
Symbol	**Prefix**	**Equivalent**
G	giga	1,000,000,000
M	mega	1,000,000
k	kilo	1000
base unit	—	1
m	milli	0.001
μ	micro	0.000001
n	nano	0.000000001

COMMON ELECTRICAL QUANTITIES		
Variable	**Name**	**Unit of Measure and Abbreviation**
E	voltage	volts — E
I	current	amperes — A
R	resistance	ohms — Ω
P	power	watts — W
P	power (apparent)	volt-amps — VA
C	capacitance	farads — F
L	inductance	henrys — H

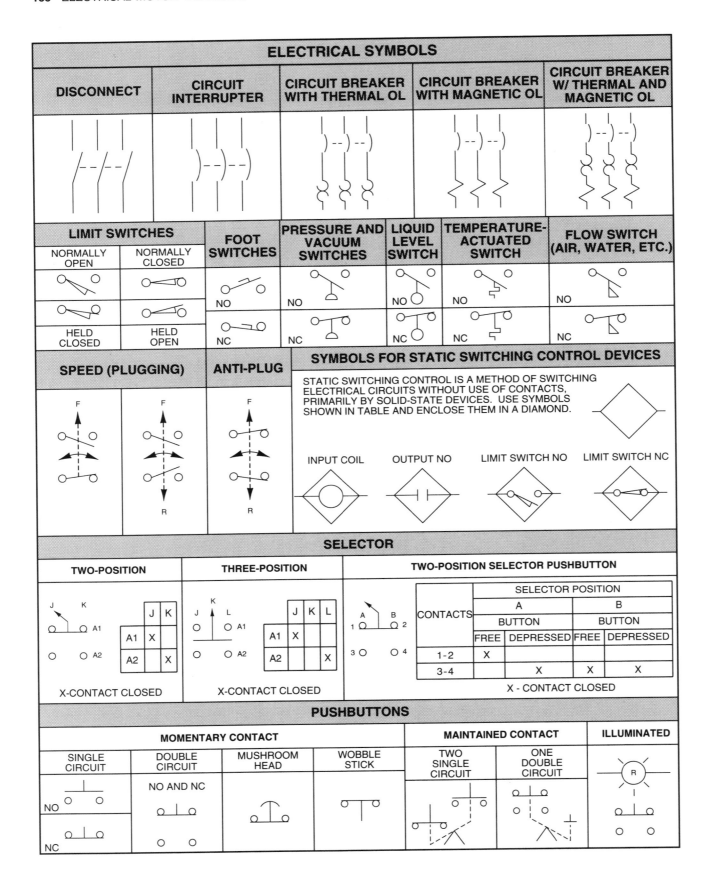

ELECTRICAL SYMBOLS

CONTACTS

INSTANT OPERATING				TIMED CONTACTS - CONTACT ACTION RETARDED AFTER COIL IS:			
WITH BLOWOUT		WITHOUT BLOWOUT		ENERGIZED		DE-ENERGIZED	
NO	NC	NO	NC	NOTC	NCTO	NOTO	NCTC

OVERLOAD RELAYS

THERMAL	MAGNETIC

SUPPLEMENTARY CONTACT SYMBOLS

SPST NO		SPST NC		SPDT		TERMS
SINGLE BREAK	DOUBLE BREAK	SINGLE BREAK	DOUBLE BREAK	SINGLE BREAK	DOUBLE BREAK	SPST SINGLE-POLE, SINGLE-THROW

DPST, 2NO		DPST, 2NC		DPDT		
SINGLE BREAK	DOUBLE BREAK	SINGLE BREAK	DOUBLE BREAK	SINGLE BREAK	DOUBLE BREAK	

SPDT
SINGLE-POLE,
DOUBLE-THROW

DPST
DOUBLE-POLE,
SINGLE-THROW

DPDT
DOUBLE-POLE,
DOUBLE-THROW

NO
NORMALLY OPEN

NC
NORMALLY CLOSED

METER (INSTRUMENT)

INDICATE TYPE BY LETTER	TO INDICATE FUNCTION OF METER OR INSTRUMENT, PLACE SPECIFIED LETTER OR LETTERS WITHIN SYMBOL.			
	AM or A	AMMETER	VA	VOLTMETER
	AH	AMPERE HOUR	VAR	VARMETER
	μA	MICROAMMETER	VARH	VARHOUR METER
	mA	MILLAMMETER	W	WATTMETER
	PF	POWER FACTOR	WH	WATTHOUR METER
	V	VOLTMETER		

PILOT LIGHTS

INDICATE COLOR BY LETTER	
NON PUSH-TO-TEST	PUSH-TO-TEST

INDUCTORS

IRON CORE

AIR CORE

COILS

DUAL-VOLTAGE MAGNET COILS		BLOWOUT COIL
HIGH-VOLTAGE	LOW-VOLTAGE	

LINK
1 2   3 4

LINKS
1 2   3 4

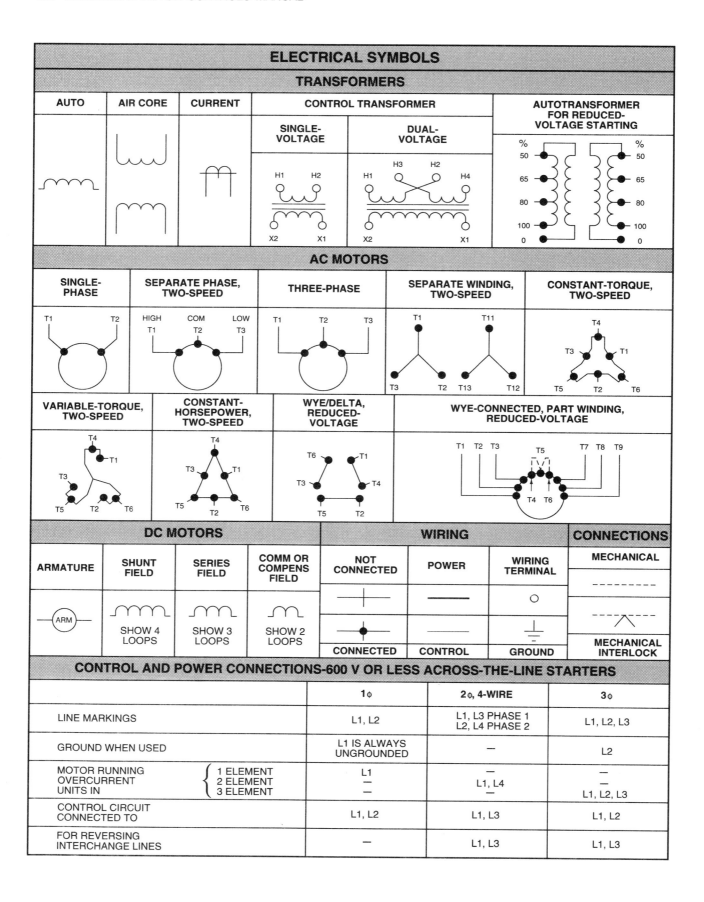

ELECTRICAL SYMBOLS

TRANSFORMERS

AUTO	AIR CORE	CURRENT	CONTROL TRANSFORMER		AUTOTRANSFORMER FOR REDUCED-VOLTAGE STARTING
			SINGLE-VOLTAGE	DUAL-VOLTAGE	

AC MOTORS

SINGLE-PHASE	SEPARATE PHASE, TWO-SPEED	THREE-PHASE	SEPARATE WINDING, TWO-SPEED	CONSTANT-TORQUE, TWO-SPEED

VARIABLE-TORQUE, TWO-SPEED	CONSTANT-HORSEPOWER, TWO-SPEED	WYE/DELTA, REDUCED-VOLTAGE	WYE-CONNECTED, PART WINDING, REDUCED-VOLTAGE

DC MOTORS / WIRING / CONNECTIONS

ARMATURE	SHUNT FIELD	SERIES FIELD	COMM OR COMPENS FIELD	NOT CONNECTED	POWER	WIRING TERMINAL	MECHANICAL
ARM	SHOW 4 LOOPS	SHOW 3 LOOPS	SHOW 2 LOOPS	CONNECTED	CONTROL	GROUND	MECHANICAL INTERLOCK

CONTROL AND POWER CONNECTIONS—600 V OR LESS ACROSS-THE-LINE STARTERS

		1 φ	2 φ, 4-WIRE	3 φ
LINE MARKINGS		L1, L2	L1, L3 PHASE 1 L2, L4 PHASE 2	L1, L2, L3
GROUND WHEN USED		L1 IS ALWAYS UNGROUNDED	—	L2
MOTOR RUNNING OVERCURRENT UNITS IN	1 ELEMENT	L1	—	—
	2 ELEMENT	—	L1, L4	—
	3 ELEMENT	—	—	L1, L2, L3
CONTROL CIRCUIT CONNECTED TO		L1, L2	L1, L3	L1, L2
FOR REVERSING INTERCHANGE LINES		—	L1, L3	L1, L3

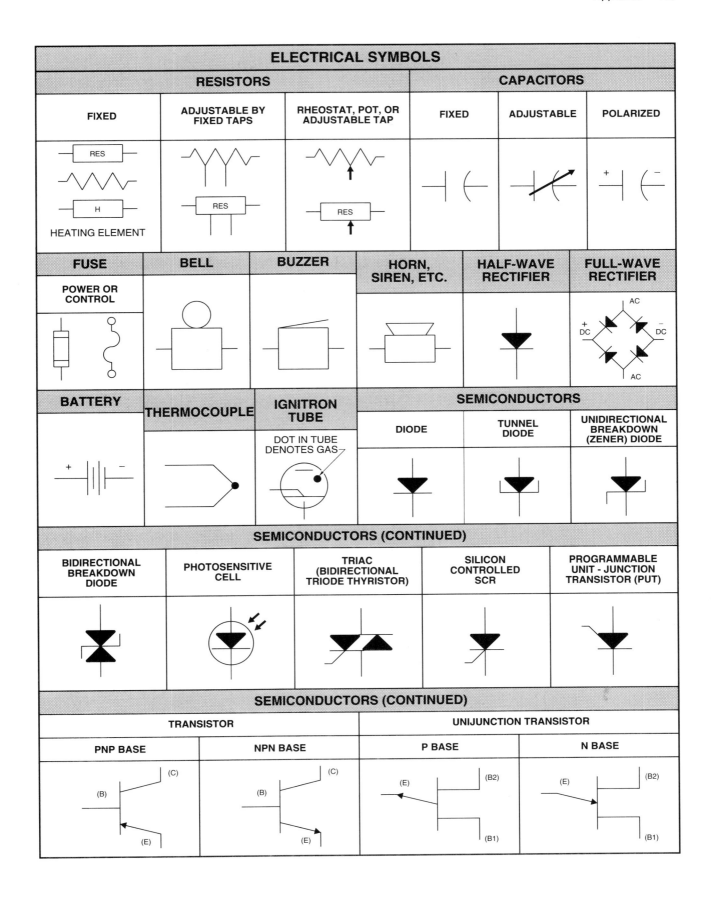

ELECTRICAL ABBREVIATIONS

Abbreviation	Term	Abbreviation	Term
AC	Alternating current	MB	Magnetic brake
ALM	Alarm	MCS	Motor circuit switch
AM	Ammeter	MEM	Memory
ARM	Armature	MTR	Motor
AU	Automatic	MN	Manual
BAT	Battery (electric)	NEG	Negative
BR	Brake relay	NEUT	Neutral
CAP	Capacitor	NC	Normally closed
CB	Circuit breaker	NO	Normally open
CEMF	Counter electromotive force	OHM	Ohmmeter
CKT	Circuit	OL	Overload relay
CONT	Control	PB	Pushbutton
CR	Control relay	PH	Phase
CRM	Control relay master	PLS	Plugging switch
CT	Current transformer	POS	Positive
D	Down	PRI	Primary switch
DB	Dynamic braking contactor or relay	PS	Pressure switch
DC	Direct current	R	Reverse
DIO	Diode	REC	Rectifier
DISC	Disconnect switch	RES	Resistor
DP	Double-pole	RH	Rheostat
DPDT	Double-pole, double-throw	S	Switch
DPST	Double-pole, single-throw	SCR	Semiconductor-control rectifier
DS	Drum switch	SEC	Secondary
DT	Double-throw	1PH	Single-phase
EMF	Electromotive force	SOC	Socket
F	Forward	SOL	Solenoid
FLS	Flow switch	SP	Single-pole
FREQ	Frequency	SPDT	Single-pole, double-throw
FS	Float switch	SPST	Single-pole, single-throw
FTS	Foot switch	SS	Selector switch
FU	Fuse	SSW	Safety switch
GEN	Generator	T	Transformer
GRD	Ground	TB	Terminal board
IC	Integrated circuit	3PH	Three-phase
INTLK	Interlock	TD	Time delay
IOL	Instantaneous overload	THS	Thermostat switch
JB	Junction box	TR	Time delay relay
LS	Limit switch	U	Up
LT	Lamp	UCL	Unclamp
M	Motor starter	UV	Undervoltage

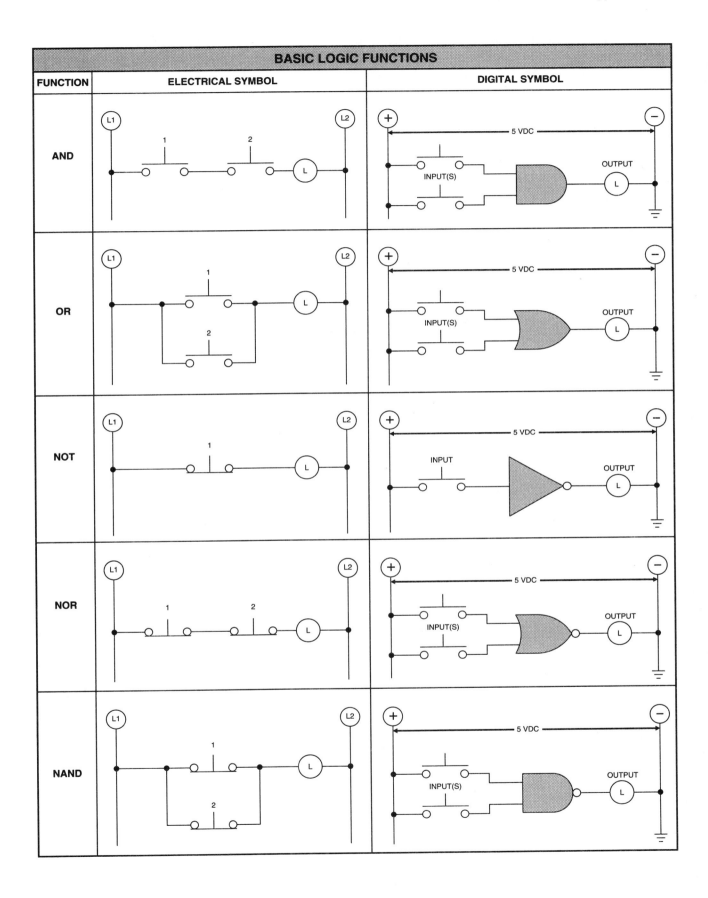

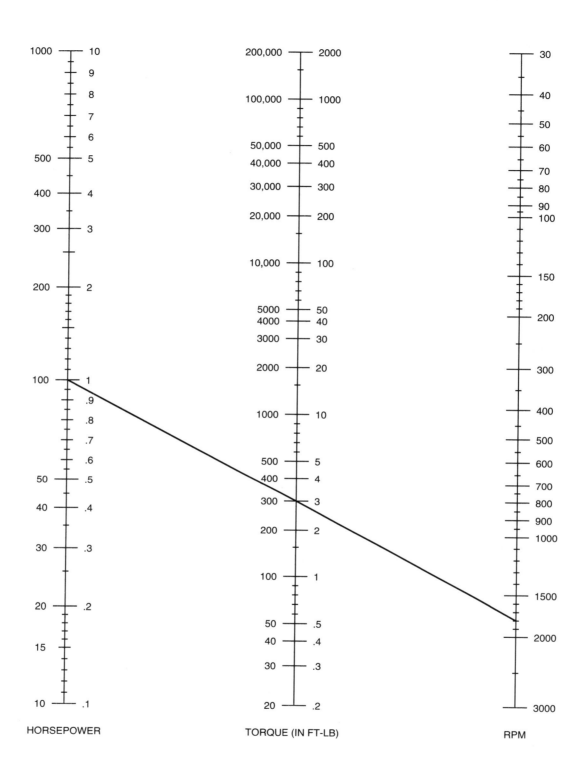

HORSEPOWER   TORQUE (IN FT-LB)   RPM

HORSEPOWER TO TORQUE CONVERSION

TORQUE FORMULA

Torque is found by applying the formula:

$$torque = \frac{HP \times 5252}{rpm}$$

where
HP = horsepower

$$5252 = constant \left(\frac{33.000 \text{ ft-lb}}{\pi \times 2} = 5252 \right)$$

rpm = revolutions per minute

For example, what is the full-load torque of a 60 HP, 240 V, 3ϕ motor, turning at 1725 rpm?

$$torque = \frac{HP \times 5252}{rpm}$$

$$torque = \frac{600 \times 5252}{1725}$$

$$torque = \frac{315,120}{1725}$$

$$torque = \textbf{182.7 ft-lb}$$

HORSEPOWER FORMULA

The horsepower of a motor, when the current and voltage are known, is found by applying the formula:

$$HP = \frac{E \times I \times E_{ff}}{746}$$

where
HP = horsepower
E = current (amps)
I = voltage (volts)
E_{ff} = efficiency

For example, what is the horsepower of a 240 V motor pulling 15 A and having 85% efficiency?

$$HP = \frac{E \times I \times E_{ff}}{746}$$

$$HP = \frac{15 \times 240 \times .85}{746}$$

$$HP = \frac{3060}{746}$$

$$HP = \textbf{4.1 HP}$$

COUPLING SELECTIONS

Coupling number	Rated torque (in-lb)	Maximum shock torque (in-lb)
10-101-A	16	45
10-102-A	36	100
10-103-A	80	220
10-104-A	132	360
10-105-A	176	480
10-106-A	240	660
10-107-A	325	900
10-108-A	525	1450

COMMON SERVICE FACTORS

Equipment	Service factor
Blowers	
Centrifugal	1.00
Vane	1.25
Compressors	
Centrifugal	1.25
Vane	1.50
Conveyors	
Uniformly loaded or fed	1.50
Heavy-duty	2.00
Elevators	
Bucket	2.00
Freight	2.25
Extruders	
Plastic	2.00
Metal	2.50
Fans	
Light-duty	1.00
Centrifugal	1.50
Machine tools	
Bending roll	2.00
Punch press	2.25
Tapping machine	3.00
Mixers	
Concrete	2.00
Drum	2.25
Paper mills	
De-barking machines	3.00
Beater and pulper	2.00
Bleacher	1.00
Dryers	2.00
Log haul	2.00
Printing presses	1.50
Pumps	
Centrifugal—general	1.00
Centrifugal—sewage	2.00
Reciprocating	2.00
Rotary	1.50
Textile	
Batchers	1.50
Dryers	1.50
Looms	1.75
Spinners	1.50
Woodworking machines	1.00

SERVICE BULLETIN

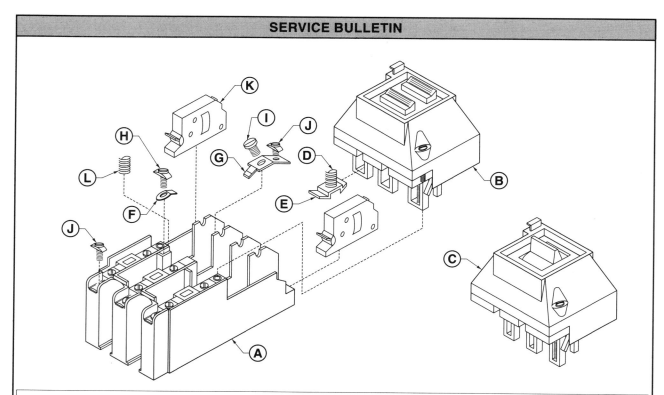

PARTS LIST

Item	Description	Part number	Size M-0 2-pole	Size M-0 3-pole	Size M-1 2-pole	Size M-1 3-pole	Size M-1P 2-pole
A	Contact Block	00401	1	1	1	1	1
B	Contact actuator—Square pushbutton						
	2-pole	01753	1	—	1	—	1
	2-pole with run-jog feature	01752	1	—	1	—	—
	3-pole	01751	—	1	—	1	—
	3-pole with run-jog feature	01750	—	1	—	1	—
C	Contact actuator—Toggle						
	2-pole	35953	1	—	1	—	1
	2-pole with run-jog feature	35952	1	—	1	—	1
	3-pole	35951	—	1	—	1	—
	3-pole with run-jog feature	35950	—	1	—	1	—
D	Contact spring	10902	1	1	—	—	—
E	Movable contact	1040	1	1	—	—	—
F	Sationary contact—Load side	1095	—	—	1	1	1
G	Sationary contact—Line side	17310	—	—	1	1	1
H	Screw assembly #6-32 x ⅜″	12080	2	3	2	3	2
I	Screw assembly #8-32 x ⅜″	14121	2	3	2	3	2
J	Wire clamp screw assembly	01850	4	6	4	6	4
K	Iternal interlock (NO)	2065	—	—	—	—	—
	Internal interlock (NC)	2066	—	—	—	—	—
L	Return spring	17901	2	2	2	2	2

1φ MOTORS AND CIRCUITS										
1	**2**		**3**	**4**	**5**				**6**	
Size of motor Table 430-148	Motor overload protection Low-peak or Fusetron®		Switch 115% minimum or HP rated or fuse holder size	Minimum size of starter	Controller termination temperature rating				Minimum size of copper wire and trade conduit	
					60°C		75°C			
HP **Amp**	Motor less than 40°C or greater than 1.15 SF (Max fuse 125%)	All other motors (Max fuse 115%)			**TW**	**THW**	**TW**	**THW**	Wire size (AWG or kcmil)	Conduit (inches)
115 V (120 V system)										
1/6 4.4	5	5	30	00	•	•	•	•	14	1/2
1/4 5.8	7	6 1/4	30	00	•	•	•	•	14	1/2
1/3 7.2	9	8	30	00	•	•	•	•	14	1/2
1/2 9.8	12	10	30	00	•	•	•	•	14	1/2
3/4 13.8	15	15	30	00	•	•	•	•	14	1/2
1 16	20	17 1/2	30	00	•	•	•	•	14	1/2
1 1/2 20	25	20	30	01	•	•	•	•	12	1/2
2 24	30	25	30	01	•	•	•	•	10	1/2
230 V (240 V system)										
1/6 2.2	2 1/2	2 1/2	30	00	•	•	•	•	14	1/2
1/4 2.9	3 1/2	3 2/10	30	00	•	•	•	•	14	1/2
1/3 3.6	4 1/2	4	30	00	•	•	•	•	14	1/2
1/2 4.9	5 6/10	5 6/10	30	00	•	•	•	•	14	1/2
3/4 6.9	8	7 1/2	30	00	•	•	•	•	14	1/2
1 8	10	9	30	00	•	•	•	•	14	1/2
1 1/2 10	12	10	30	0	•	•	•	•	14	1/2
2 12	15	12	30	0	•	•	•	•	14	1/2
3 17	20	17 1/2	30	1	•	•	•	•	12	1/2
5 28	35	30	60	2		•			8	3/4
					•		•		8	1/2
								•	10	1/2
7 1/2 40	50	45	60	2	•	•	•		6	3/4
								•	8	3/4
									8	1/2
10 50	60	50	60	3	•	•	•		4	1
									4	3/4
								•	6	3/4

3φ, 230 V MOTORS AND CIRCUITS — 240 V SYSTEM											
1 Size of motor Table 430-150		**2** Motor overload protection Low-peak or Fusetron®		**3**	**4**	**5** Controller termination temperature rating				**6** Minimum size of copper wire and trade conduit	
						60°C		75°C			
		Motor less than 40°C or greater than 1.15 SF (Max fuse 125%)	All other motors (Max fuse 115%)	Switch 115% minimum or HP rated or fuse holder size	Minimum size of starter					Wire size (AWG or kcmil)	Conduit (inches)
HP	Amp					TW	THW	TW	THW		
½	2	2½	2¼	30	00	•	•	•	•	14	½
¾	2.8	3½	3²/₁₀	30	00	•	•	•	•	14	½
1	3.6	4½	4	30	00	•	•	•	•	14	½
1½	5.2	6¼	5⁶/₁₀	30	00	•	•	•	•	14	½
2	6.8	8	7½	30	0	•	•	•	•	14	½
3	9.6	12	10	30	0	•	•	•	•	14	½
5	15.2	17½	17½	30	1	•	•	•	•	14	½
7½	22	25	25	30	1	•	•	•	•	10	½
10	28	35	30*	60	2	•	•	•		8	¾
										8	½
									•	10	½
15	42	50	45	60	2	•	•	•	•	6	1
										6	¾
20	54	60*	60*	100	3	•	•	•	•	4	1
25	68	80	75	100	3	•	•			3	1¼
								•		3	1
									•	4	1
30	80	100	90	100	3	•	•	•		1	1¼
									•	3	1¼
										3	1
40	104	125	110	200	4	•	•	•		2/0	1½
									•	1	1¼
50	130	150	150	200	4	•	•	•		3/0	2
										3/0	1½
									•	2/0	1½
75	192	225	200*	400	5	•	•	•		300	2½
										300	2
									•	250	2½
										250	2
100	248	300	250	400	5	•	•	•		500	3
									•	350	2½
150	360	450	400*	600	6	•	•	•		300-2φ*	2-2½*
										300-2φ*	2-2*
									•	4/0-2φ*	2-2*

*Fuse reducers required.

1		2		3	4	5				6	
3φ, 460 V MOTORS AND CIRCUITS — 480 V SYSTEM											
Size of motor Table 430-150		**Motor overload protection** Low-peak or Fusetron®		**Switch 115% minimum or HP rated or fuse holder size**	**Minimum size of starter**	**Controller termination temperature rating**				**Minimum size of copper wire and trade conduit**	
						60°C		75°C			
		Motor less than 40°C or greater than 1.15 SF (Max fuse 125%)	All other motors (Max fuse 115%)			TW	THW	TW	THW	Wire size (AWG or kcmil)	Conduit (inches)
HP	Amp										
½	1	1¼	1⅛	30	00	●	●	●	●	14	½
¾	1.4	1⁶⁄₁₀	1⁶⁄₁₀	30	00	●	●	●	●	14	½
1	1.8	2¼	2	30	00	●	●	●	●	14	½
1½	2.6	3²⁄₁₀	2⁶⁄₁₀	30	00	●	●	●	●	14	½
2	3.4	4	3½	30	00	●	●	●	●	14	½
3	4.8	5⁶⁄₁₀	5	30	0	●	●	●	●	14	½
5	7.6	9	8	30	0	●	●	●	●	14	½
7½	11	12	12	30	1	●	●	●	●	14	½
10	14	17½	15	30	1	●	●	●	●	14	½
15	21	25	20	30	2	●	●	●	●	10	½
20	27	30*	30*	60	2	●	●	●		8	¾
										8	½
									●	10	½
25	34	40	35	60	2	●	●	●		6	1
										6	¾
									●	8	¾
										8	½
30	40	50	45	60	3	●	●	●		6	1
										6	¾
									●	8	¾
										8	½
40	52	60*	60*	100	3	●	●	●		4	1
										6	¾
									●	6	1
50	65	80	70	100	3	●	●	●		3	1¼
										3	1
									●	4	1
60	77	90	80	100	4	●	●	●		1	1¼
										3	1
									●	3	1¼
75	96	110	110	200	4	●	●	●		1/0	1½
										1/0	1¼
									●	1	1¼

*Fuse reducers required.

continued

continued

3φ, 460 V MOTORS AND CIRCUITS — 480 V SYSTEM											
1		**2**		**3**	**4**	**5**				**6**	
Size of motor Table 430-150		Motor overload protection Low-peak or Fusetron®		Switch 115% minimum or HP rated or fuse holder size	Minimum size of starter	Controller termination temperature rating				Minimum size of copper wire and trade conduit	
						60°C		75°C			
		Motor less than 40°C or greater than 1.15 SF (Max fuse 125%)	All other motors (Max fuse 115%)							Wire size (AWG or kcmil)	Conduit (inches)
HP	Amp					TW	THW	TW	THW		
100	124	150	125	200	4	•	•	•		3/0	2
										3/0	1½
									•	2/0	1½
125	156	175	175	200	5	•	•	•		4/0	2
									•	3/0	2
										3/0	1½
150	180	225	200*	400	5	•	•	•		300	2½
										300	2
									•	4/0	2
200	240	300	250	400	5	•	•	•		500	3
									•	350	2½
250	302	350	325	400	6	•	•	•		4/0-2φ*	2-2*
									•	3/0-2φ*	2-2*
										3/0-2φ*	1½*
300	361	450	400*	600	6	•	•	•		300-2φ*	2-1½*
										300-2φ*	2-2*
									•	4/0-2φ*	2-2*

*Fuse reducers required.

DC MOTORS AND CIRCUITS

HP	Amp	Motor less than 40°C or greater than 1.15 SF (Max fuse 125%)	All other motors (Max fuse 115%)	Switch 115% minimum or HP rated or fuse holder size	Minimum size of starter	TW (60°C)	THW (60°C)	TW (75°C)	THW (75°C)	Wire size (AWG or kcmil)	Conduit (in.)
90 V											
¼	4.0	5	4½	30	0	•	•	•	•	14	½
⅓	5.2	6¼	5⁶⁄₁₀	30	0	•	•	•	•	14	½
½	6.8	8	7½	30	0	•	•	•	•	14	½
¾	9.6	12	10	30	0	•	•	•	•	14	½
1	12.2	15	12	30	0	•	•	•	•	14	½
120 V											
¼	3.1	3½	3½	30	0	•	•	•	•	14	½
⅓	4.1	5	4½	30	0	•	•	•	•	14	½
½	5.4	6¼	6	30	0	•	•	•	•	14	½
¾	7.6	9	8	30	0	•	•	•	•	14	½
1	9.5	10	10	30	0	•	•	•	•	14	½
1½	13.2	15	15	30	1	•	•	•	•	14	½
2	17	20	17½	30	1	•	•	•	•	12	½
5	40	50	45	60	2	•	•	•		6	¾
									•	8	¾
										8	½
10	76	90	80	100	3	•	•	•		2	1
									•	3	1
180 V											
¼	2	2½	2¼	30	0	•	•	•	•	14	½
⅓	2.6	3²⁄₁₀	2⁸⁄₁₀	30	0	•	•	•	•	14	½
½	3.4	4	3½	30	0	•	•	•	•	14	½
¾	4.8	6	5	30	0	•	•	•	•	14	½
1	6.1	7½	7	30	0	•	•	•	•	14	½
1½	8.3	10	9	30	1	•	•	•	•	14	½
2	10.8	12	12	30	1	•	•	•	•	14	½
3	16	20	17½	30	1	•	•	•	•	12	½
5	27	30*	30*	60	1	•		•		8	½
							•			8	¾
									•	10	½

Column headers:
1 — Size of motor, Table 430-147 (HP, Amp)
2 — Motor overload protection, Dual-element fuse
3 — Switch 115% minimum or HP rated or fuse holder size
4 — Minimum size of starter
5 — Controller termination temperature rating (60°C, 75°C)
6 — Minimum size of copper wire and trade conduit

*Fuse reducers required.

FLUID POWER SYMBOLS

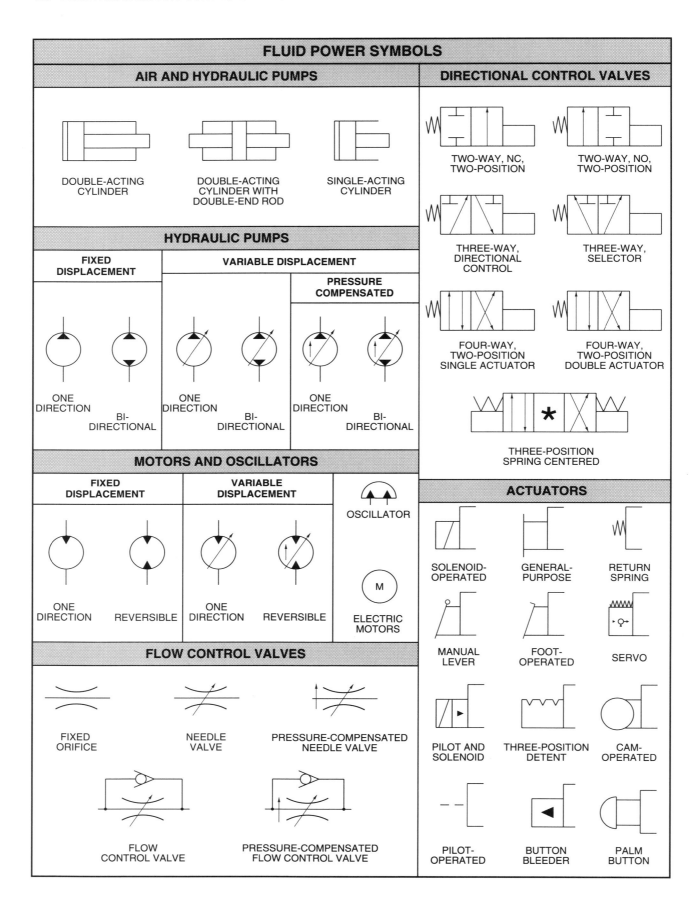

AIR AND HYDRAULIC PUMPS

DOUBLE-ACTING
CYLINDER

DOUBLE-ACTING
CYLINDER WITH
DOUBLE-END ROD

SINGLE-ACTING
CYLINDER

HYDRAULIC PUMPS

**FIXED
DISPLACEMENT**

VARIABLE DISPLACEMENT

**PRESSURE
COMPENSATED**

ONE
DIRECTION

BI-
DIRECTIONAL

ONE
DIRECTION

BI-
DIRECTIONAL

ONE
DIRECTION

BI-
DIRECTIONAL

MOTORS AND OSCILLATORS

**FIXED
DISPLACEMENT**

**VARIABLE
DISPLACEMENT**

OSCILLATOR

ONE
DIRECTION

REVERSIBLE

ONE
DIRECTION

REVERSIBLE

M

ELECTRIC
MOTORS

FLOW CONTROL VALVES

FIXED
ORIFICE

NEEDLE
VALVE

PRESSURE-COMPENSATED
NEEDLE VALVE

FLOW
CONTROL VALVE

PRESSURE-COMPENSATED
FLOW CONTROL VALVE

DIRECTIONAL CONTROL VALVES

TWO-WAY, NC,
TWO-POSITION

TWO-WAY, NO,
TWO-POSITION

THREE-WAY,
DIRECTIONAL
CONTROL

THREE-WAY,
SELECTOR

FOUR-WAY,
TWO-POSITION
SINGLE ACTUATOR

FOUR-WAY,
TWO-POSITION
DOUBLE ACTUATOR

★

THREE-POSITION
SPRING CENTERED

ACTUATORS

SOLENOID-
OPERATED

GENERAL-
PURPOSE

RETURN
SPRING

MANUAL
LEVER

FOOT-
OPERATED

SERVO

PILOT AND
SOLENOID

THREE-POSITION
DETENT

CAM-
OPERATED

PILOT-
OPERATED

BUTTON
BLEEDER

PALM
BUTTON

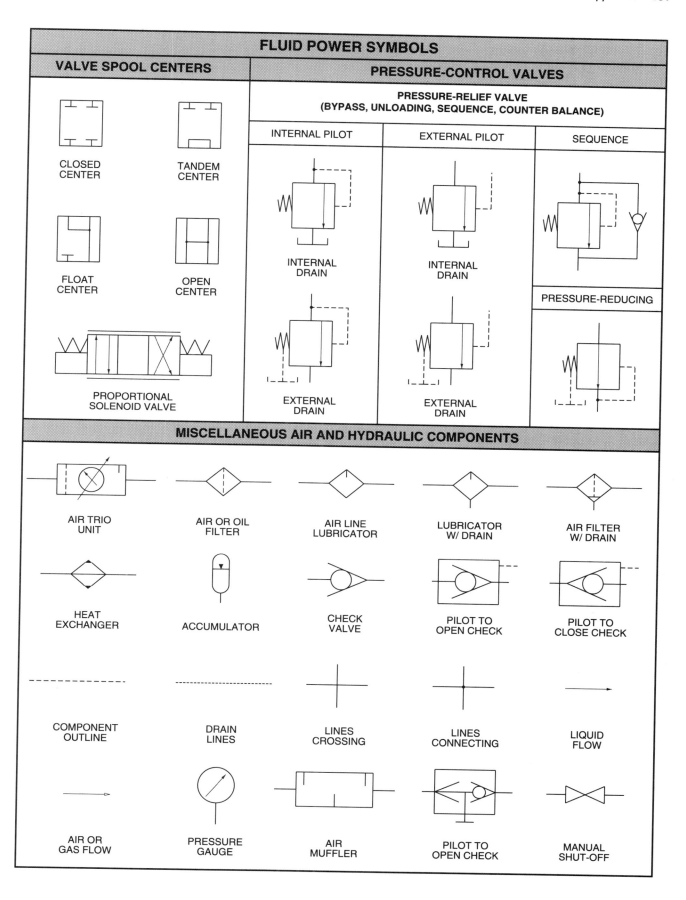

FLUID POWER SYMBOLS

VALVE SPOOL CENTERS

CLOSED CENTER

TANDEM CENTER

FLOAT CENTER

OPEN CENTER

PROPORTIONAL SOLENOID VALVE

PRESSURE-CONTROL VALVES

PRESSURE-RELIEF VALVE
(BYPASS, UNLOADING, SEQUENCE, COUNTER BALANCE)

INTERNAL PILOT	EXTERNAL PILOT	SEQUENCE
INTERNAL DRAIN	INTERNAL DRAIN	
EXTERNAL DRAIN	EXTERNAL DRAIN	PRESSURE-REDUCING

MISCELLANEOUS AIR AND HYDRAULIC COMPONENTS

AIR TRIO UNIT

AIR OR OIL FILTER

AIR LINE LUBRICATOR

LUBRICATOR W/ DRAIN

AIR FILTER W/ DRAIN

HEAT EXCHANGER

ACCUMULATOR

CHECK VALVE

PILOT TO OPEN CHECK

PILOT TO CLOSE CHECK

COMPONENT OUTLINE

DRAIN LINES

LINES CROSSING

LINES CONNECTING

LIQUID FLOW

AIR OR GAS FLOW

PRESSURE GAUGE

AIR MUFFLER

PILOT TO OPEN CHECK

MANUAL SHUT-OFF

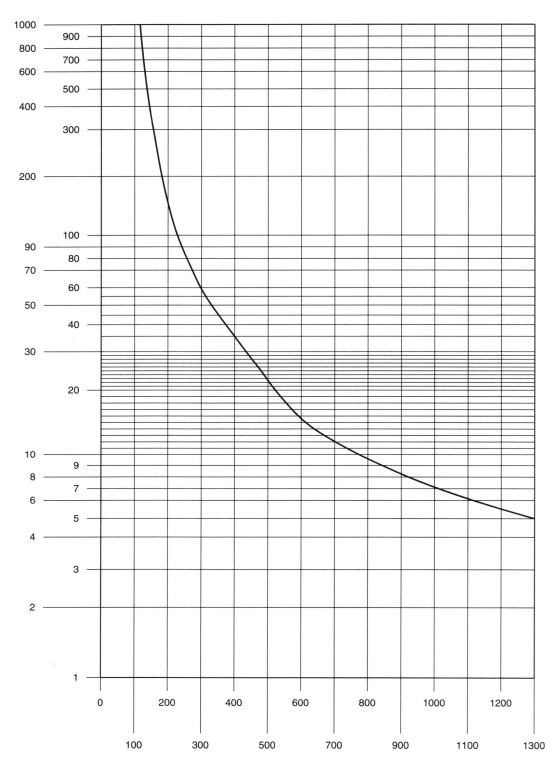

RATED CURRENT (%)

HEATER TRIP CHARACTERISTICS

HEATER SELECTIONS					
Heater number	Full load current (A)				
	Size 0	Size 1	Size 2	Size 3	Size 4
10	.20	.20	—	—	—
11	.22	.22	—	—	—
12	.24	.24	—	—	—
13	.27	.27	—	—	—
14	.30	.30	—	—	—
15	.34	.34	—	—	—
16	.37	.37	—	—	—
17	.41	.41	—	—	—
18	.45	.45	—	—	—
19	.49	.49	—	—	—
20	.54	.54	—	—	—
21	.59	.59	—	—	—
22	.65	.65	—	—	—
23	.71	.71	—	—	—
24	.78	.78	—	—	—
25	.85	.85	—	—	—
26	.93	.93	—	—	—
27	1.02	1.02	—	—	—
28	1.12	1.12	—	—	—
29	1.22	1.22	—	—	—
30	1.34	1.34	—	—	—
31	1.48	1.48	—	—	—
32	1.62	1.62	—	—	—
33	1.78	1.78	—	—	—
34	1.96	1.96	—	—	—
35	2.15	2.15	—	—	—
36	2.37	2.37	—	—	—
37	2.60	2.60	—	—	—
38	2.86	2.86	—	—	—
39	3.14	3.14	—	—	—
40	3.45	3.45	—	—	—
41	3.79	3.79	—	—	—
42	4.17	4.17	—	—	—
43	4.58	4.58	—	—	—
44	5.03	5.03	—	—	—
45	5.53	5.53	—	—	—
46	6.08	6.08	—	—	—

HEATER SELECTIONS					
Heater number	Full load current (A)				
	Size 0	Size 1	Size 2	Size 3	Size 4
47	6.68	6.68	—	—	—
48	7.21	7.21	—	—	—
49	7.81	7.81	7.89	—	—
50	8.46	8.46	8.57	—	—
51	9.35	9.35	9.32	—	—
52	10.00	10.00	10.1	—	—
53	10.7	10.7	11.0	12.2	—
54	11.7	11.7	12.0	13.3	—
55	12.6	12.6	12.9	14.3	—
56	13.9	13.9	14.1	15.6	—
57	15.1	15.1	15.5	17.2	—
58	16.5	16.5	16.9	18.7	—
59	18.0	18.0	18.5	20.5	—
60	—	19.2	20.3	22.5	23.8
61	—	20.4	21.8	24.3	25.7
62	—	21.7	23.5	26.2	27.8
63	—	23.1	25.3	28.3	30.0
64	—	24.6	27.2	30.5	32.5
65	—	26.2	29.3	33.0	35.0
66	—	27.8	31.5	36.0	38.0
67	—	—	33.5	39.0	41.0
68	—	—	36.0	42.0	44.5
69	—	—	38.5	45.5	48.5
70	—	—	41.0	49.5	52
71	—	—	43.0	53	57
72	—	—	46.0	58	61
73	—	—	—	63	67
74	—	—	—	68	72
75	—	—	—	73	77
76	—	—	—	78	84
77	—	—	—	83	91
78	—	—	—	88	97
79	—	—	—	—	103
80	—	—	—	—	111
81	—	—	—	—	119
82	—	—	—	—	127
83	—	—	—	—	133

FULL-LOAD CURRENTS — 3φ, AC INDUCTION MOTORS

Motor rating (HP)	Current (A)			
	208 V	230 V	460 V	575 V
¼	1.11	.96	.48	.38
⅓	1.34	1.18	.59	.47
½	2.2	2.0	1.0	.8
¾	3.1	2.8	1.4	1.1
1	4.0	3.6	1.8	1.4
1½	5.7	5.2	2.6	2.1
2	7.5	6.8	3.4	2.7
3	10.6	9.6	4.8	3.9
5	16.7	15.2	7.6	6.1
7½	24.0	22.0	11.0	9.0
10	31.0	28.0	14.0	11.0
15	46.0	42.0	21.0	17.0
20	59	54	27	22
25	75	68	34	27
30	88	80	40	32
40	114	104	52	41
50	143	130	65	52
60	169	154	77	62
75	211	192	96	77
100	273	248	124	99
125	343	312	156	125
150	396	360	180	144
200	—	480	240	192
250	—	602	301	242
300	—	—	362	288
350	—	—	413	337
400	—	—	477	382
500	—	—	590	472

FULL-LOAD CURRENTS — 1φ, AC MOTORS

Motor rating (HP)	Current (A)	
	115 V	230 V
⅙	4.4	2.2
¼	5.8	2.9
⅓	7.2	3.6
½	9.8	4.9
¾	13.8	6.9
1	16	8
1½	20	10
2	24	12
3	34	17
5	56	28
7½	80	40
10	100	50

FULL-LOAD CURRENTS — DC MOTORS

Motor rating (HP)	Current (A)	
	120 V	240 V
¼	3.1	1.6
⅓	4.1	2.0
½	5.4	2.7
¾	7.6	3.8
1	9.5	4.7
1½	13.2	6.6
2	17	8.5
3	25	12.2
5	40	20
7½	48	29

CONTROL RATINGS

Size	Load (V)	Maximum HP Normal duty 1φ	3φ	Plugging & jogging duty 1φ	3φ	Cont amps	Service limit amps	Tungsten & ballast type lamp amps 480 V max	Resistance heating (kW) 1φ	3φ	Transformer switching 50–60 Hz kVA rating inrush peak time Continuous amps 20 times 1φ	3φ	20-40 times 1φ	3φ	Capacitor kVA switching rating 3φ kVAR
00	115	½	—	—	—	9	11	—	1.15	2.0	—	—	—	—	—
	200	—	1½	—	—	9	11	—	2.0	3.46	—	—	—	—	—
	230	1	1½	—	—	9	11	—	2.3	4.0	—	—	—	—	—
	380	—	1½	—	—	9	11	—	—	6.5	—	—	—	—	—
	460	—	2	—	—	9	11	—	4.6	8.0	—	—	—	—	—
	575	—	2	—	—	9	11	—	5.8	10.0	—	—	—	—	—
0	115	1	—	½	—	18	21	20	2.3	4.0	0.6	—	0.3	—	—
	200	—	3	—	1½	18	21	20	4.0	6.92	—	1.8	—	0.9	—
	230	2	3	1	1½	18	21	20	4.6	8.0	1.2	2.1	0.6	1.0	—
	380	—	5	—	1½	18	21	20	—	13.1	—	—	—	—	—
	460	—	5	—	2	18	21	20	9.2	15.9	2.4	4.2	1.2	2.1	—
	575	—	5	—	2	18	21	—	11.5	19.9	3.0	5.2	1.5	2.6	—
1	115	2	—	1	—	27	32	30	3.5	6.0	1.2	—	0.6	—	—
	200	—	7½	—	3	27	32	30	6	10.4	—	3.6	—	1.8	—
	230	3	7½	2	3	27	32	30	6.9	11.9	2.4	4.3	1.2	2.1	—
	380	—	10	—	5	27	32	30	—	19.7	—	—	—	—	—
	460	—	10	—	5	27	32	30	13.8	23.9	4.9	8.5	2.5	4.3	—
	575	—	10	—	5	27	32	—	17.3	29.8	6.2	11.0	3.1	5.3	—
1P	115	3	—	1½	—	35	42	45	5.8	—	—	—	—	—	—
	230	5	—	3	—	35	42	45	11.5	—	—	—	—	—	—
1¾	115	—	—	—	—	40	40	45	5.8	9.9	1.6	—	0.8	—	—
	200	—	10	—	5	40	40	45	10	17.3	—	4.9	—	2.4	—
	230	—	10	—	5	40	40	45	11.5	19.9	3.2	5.75	1.6	2.8	—
	380	—	15	—	7½	40	40	45	—	32.9	—	—	—	—	—
	460	—	15	—	7½	40	40	45	23	39.8	6.6	11.2	3.3	5.7	—
	575	—	15	—	7½	40	40	—	28.8	49.7	8.1	14.5	4.1	7.1	—
2	115	3	—	2	—	45	52	60	8.1	13.9	2.1	—	1.0	—	—
	200	—	10	—	7½	45	52	60	14	24.2	—	6.3	—	3.1	—
	230	7½	15	5	10	45	52	60	16.1	27.8	4.1	7.2	2.1	3.6	8
	380	—	25	—	15	45	52	60	—	46.0	—	—	—	—	—
	460	—	25	—	15	45	52	60	32.2	55.7	8.3	14	4.2	7.2	16
	575	—	25	—	15	45	52	—	40.3	69.6	10.0	18	5.2	8.9	20
2½	115	5	—	—	—	60	65	75	10.4	17.9	3.1	—	1.5	—	—
	200	—	15	—	10	60	65	75	18	31.1	—	9.1	—	4.6	—
	230	10	20	—	15	60	65	75	20.7	35.8	6.1	10.6	3.1	5.3	17.5
	380	—	30	—	20	60	65	75	—	59.2	—	—	—	—	—
	460	—	30	—	20	60	65	75	41.4	71.6	12	21	6.1	10.6	34.5
	575	—	30	—	20	60	65	—	51.8	89.5	15	26.5	7.6	13.4	43.5

continued

continued

Size	Load (V)	Normal duty 1φ	Normal duty 3φ	Plugging & jogging duty 1φ	Plugging & jogging duty 3φ	Cont. amps	Service limit amps	Tungsten & ballast type lamp amps 480 V max	Resistance heating (kW) 1φ	Resistance heating (kW) 3φ	20 times 1φ	20 times 3φ	20-40 times 1φ	20-40 times 3φ	Capacitor kVA switching rating 3φ kVAR
3	115	7½	—	—	—	90	104	100	14.4	24.8	4.1	—	2.0	—	—
	200	—	25	—	15	90	104	100	25	43.3	—	12	—	6.1	—
	230	15	30	—	20	90	104	100	28.8	50.0	8.1	14	4.1	7.0	27
	380	—	50	—	30	90	104	100	—	82.2	—	—	—	—	—
	460	—	50	—	30	90	104	100	57.5	99.4	16	28	8.1	14	53
	575	—	50	—	30	90	104	—	71.9	124	20	35	10	18	67
3½	115	—	—	—	—	115	125	150	18.4	31.8	—	—	—	—	—
	200	—	30	—	20	115	125	150	32	55.4	—	16	—	8	—
	230	—	60	—	25	115	125	150	36.8	63.7	11	18.5	5.4	9.5	33.5
	380	—	60	—	30	115	125	150	—	105	—	—	—	—	—
	460	—	75	—	40	115	125	150	73.6	127	21.5	37.5	11.0	18.5	66.5
	575	—	75	—	40	115	125	—	92	159	37	47	13.5	23.5	83.5
4	200	—	40	—	25	135	156	200	39	67.5	—	20	—	10	—
	230	—	50	—	30	135	156	200	44.9	77.6	14	23	6.8	12	40
	380	—	75	—	50	135	156	200	—	128	—	—	—	—	—
	460	—	100	—	60	135	156	200	89.7	155	27	47	14	23	80
	575	—	100	—	60	135	156	—	112	194	34	59	17	29	100
4½	200	—	50	—	30	210	225	250	53	91.7	—	30.5	—	15	—
	230	—	75	—	40	210	225	250	60.9	105	20.5	35	10.4	18	60
	380	—	100	—	75	210	225	250	—	174	—	—	—	—	—
	460	—	150	—	100	210	225	250	122	211	40.5	70.5	20.5	35	120
	575	—	150	—	100	210	225	—	152	264	51	88	25.5	44	150